Evolution and Creation

Different Questions, One Path

Kendir Ramiz

Evolution and Creation Different Questions, One Path

The Endless Question

Imagine standing alone beneath a canopy of stars on a clear, moonless night. The immensity of the cosmos stretches out above you, infinite and incomprehensible. You feel the weight of your own existence, small and fleeting against the backdrop of eternity. In this moment, a question rises unbidden in your mind—a question as old as humanity itself: Who am I?

This question has no single answer, yet it persists, echoing across the millennia. It is the thread that connects the first humans who gazed at the heavens with wonder, the great thinkers who mapped the workings of the natural world, and the modern scientists who peer into the depths of atoms and galaxies. It is the question that drives exploration, innovation, and creation. And it is the question that lies at the heart of this book.

To understand who we are, we must embark on a journey that spans billions of years and countless lifetimes. It is a journey through the story of evolution, the myths of creation, and the interplay between the two. It is a journey that will take us from the swirling chaos of the universe's birth to the rise of life on Earth, from the first sparks of human consciousness to the technological revolutions that now shape our world. Along the way, we will confront mysteries that defy simple explanation, moments of transformation that altered the course of life forever, and the profound connections that bind us to all that has come before.

But this is not a story of facts alone. This is a story about meaning—about the search for purpose in a world that can seem both wondrous and indifferent. It is about the tension between what we can know and what remains beyond our grasp. It is about the way science and faith, often seen as adversaries, might instead be partners in a shared quest to understand the universe and our place within it.

The answers we uncover may not be comforting. Evolution is not a narrative of progress or perfection; it is a tale of survival and change, shaped by forces beyond our control. And yet, within this randomness, there is beauty. Life, in all its forms, is a testament to resilience and creativity. From the tiniest microbe to the towering redwood, from the delicate wings of a butterfly to the complex workings of the human brain, each is a product of billions of years of adaptation and innovation. And at the center of this intricate web of life stands Homo sapiens—a species both extraordinary and deeply flawed, capable of wonder and destruction in equal measure.

To explore the story of life is to explore ourselves. The same forces that shaped the first living cells have shaped the structure of your DNA. The same cosmic events that forged the elements of the Earth have left their imprint on your body. You are, quite literally, a

Evolution and Creation Different Questions, One Path

product of the universe—an ephemeral yet integral part of a vast and ongoing story.

Yet, for all our knowledge, gaps remain. There are leaps in the fossil record that defy explanation, evolutionary mysteries that tantalize and perplex, and questions about consciousness and morality that science has yet to answer. These gaps are not failures but opportunities—spaces where imagination and inquiry meet, where the boundaries of what we know are pushed ever further.

And then there is the future. Evolution is not a relic of the past; it is a process that continues, shaping life in ways both visible and invisible. But as Homo sapiens has learned to manipulate its environment, it has also begun to shape its own evolutionary path. What does it mean to be human in an era of genetic engineering, artificial intelligence, and ecological crisis? How do we navigate a world of our own making while remaining connected to the natural order that sustains us?

This book is an invitation—to wonder, to question, and to explore. It is not a search for definitive answers but a celebration of the questions that define us. It is a call to look up at the stars, down at the Earth, and within ourselves to trace the threads of a story that is as ancient as the cosmos and as immediate as the beating of your heart.

Kendir Ramiz

The Difference Between Laws and Theories in Evolution

Science thrives on clarity, yet its language often seems to invite confusion. When people discuss evolution, they frequently conflate "laws" and "theories," treating them as interchangeable or, worse, assuming that one is inherently superior to the other. This misunderstanding is not merely semantic—it distorts the very essence of what science aims to achieve. Evolution, as a concept, is anchored in both laws and theories, each playing a distinct role in unraveling the complexities of life.

A law in science is a universal statement about how something works under specific conditions. It describes phenomena with precision and reliability, much like the law of gravity dictates the behavior of falling objects. Laws are the "what" of science: observable patterns that remain consistent across time and space. Evolutionary laws, such as the inheritance of genetic material or the principles of natural selection, define the mechanisms that govern the changes we see in living organisms.

Theories, on the other hand, tackle the "how" and "why." They are frameworks that seek to explain the mechanisms behind the laws. Theories are dynamic, evolving as new evidence emerges. Charles Darwin's theory of natural selection is one of the most celebrated examples in science. It provides a narrative for how species adapt and survive within the confines of the evolutionary laws that govern them.

Evolution and Creation Different Questions, One Path

Understanding this distinction is crucial. A law does not "graduate" into a theory, nor does a theory evolve into a law. They are not steps on a hierarchical ladder but complementary tools. Together, they form a mosaic that reveals the intricacies of the natural world.

In evolutionary biology, the laws are immutable—genetic mutations, the passage of traits through heredity, and the survival of organisms best suited to their environments remain constants. Theories, however, are ever-shifting landscapes. While Darwin's natural selection theory forms the bedrock of evolutionary thought, modern science has enriched it with new perspectives, from genetic drift to epigenetics. Each addition refines the narrative without undermining the foundational truths.

This distinction illuminates another vital facet of science: its reliance on observation. The scientific method demands evidence, repeatability, and falsifiability. Evolutionary biology excels in this regard. Fossil records, genetic analyses, and controlled laboratory experiments provide a wealth of data that underpin its laws and validate its theories. Richard Lenski's long-term E. coli experiment, for instance, has demonstrated evolution in real time, as generations of bacteria adapted to new environments over decades of study. Such experiments transform the abstract into the tangible, proving that evolution is not merely a theory in the colloquial sense but a demonstrable reality.

Yet, the scientific method also acknowledges its limitations. It does not claim to answer every question definitively but instead offers a process for continual inquiry. Evolutionary theories adapt and expand precisely because science is designed to accommodate uncertainty and challenge. This adaptability is not a weakness but a testament to its strength—a recognition that knowledge is never static.

What often complicates discussions about evolution is the perception that science and belief systems are at odds. Critics of evolution sometimes argue that its theories are speculative and lack the certainty of religious truths. This perspective misunderstands the purpose of science. It is not a rival to faith but a method of understanding the natural world. Science asks "how" and "what," while faith often asks "why." These are not opposing questions but complementary ones.

Moreover, evolutionary science invites a sense of wonder that is not unlike spiritual reflection. The intricate mechanisms of genetic inheritance, the elegant simplicity of natural selection, and the resilience of life in the face of constant change—all these elements evoke awe. Recognizing the laws that govern evolution does not diminish the mystery of existence; it deepens it. To study evolution is to trace the threads of a tapestry that spans billions of years, revealing a narrative of persistence, adaptation, and interconnectedness.

This perspective shifts the conversation. Evolution is not a battle between science and belief but an opportunity to

Evolution and Creation Different Questions, One Path

explore the harmony between them. The laws of evolution do not negate the idea of creation; they articulate the processes through which creation unfolds. Whether one views these processes as the hand of a divine creator or the result of natural phenomena, the underlying principles remain the same.

Thus, the laws and theories of evolution are not mere academic constructs; they are lenses through which we understand our place in the universe. They challenge us to look beyond the surface, to question assumptions, and to embrace the complexity of life. In doing so, they open the door to deeper inquiry—not just about where we come from, but about who we are and where we might go.

Kendir Ramiz

The Law of Evolution: Observable Processes of Natural Selection, Mutation, and Genetic Change

Evolution, at its core, is a process—fluid, dynamic, and unrelenting. To grasp its essence, we must first untangle the specific mechanisms that drive it, mechanisms as fundamental to life as the laws of gravity are to the motion of the planets. These mechanisms—natural selection, mutation, and genetic change—form the backbone of what we understand as the "law of evolution." Unlike theories, which offer explanations, laws describe patterns: immutable, observable, and universal truths about how the natural world functions.

The law of evolution is not an abstract construct confined to academic papers or distant epochs. It is happening around us, within us, and even because of us. It reveals itself in the antibiotic resistance of bacteria, the shifting beak shapes of Darwin's finches, and the gradual adaptation of plants to rising global temperatures. Evolution is not a relic of the past; it is an

Evolution and Creation Different Questions, One Path

ongoing phenomenon, as current and immediate as the air we breathe.

Natural selection, often misunderstood as a principle of brute competition, is in fact a subtle and elegant process. Organisms do not "struggle" in a gladiatorial sense; they persist—or fail to persist—based on their fit within an environment. This "fitness" is not about strength or intelligence but about suitability. A species' survival hinges on its ability to align with the ecological niche it occupies, a process shaped over countless generations. Charles Darwin described this phenomenon as the "preservation of favored races," though modern understanding has clarified the nuances and implications of his work.

Consider the Galápagos finches that so famously inspired Darwin. The variation in their beak shapes—long and slender for probing cacti, short and robust for cracking seeds—is not a random accident. Over generations, environmental pressures have sculpted these adaptations, favoring certain traits while others faded into evolutionary obscurity. This is natural selection in action: the environment as both a sculptor and a judge, rewarding traits that align with survival and penalizing those that do not.

But natural selection cannot act alone. Mutation is its indispensable partner, the wellspring of diversity. Mutations are often painted as errors in the genetic code, as though life were a manuscript and any deviation from the script an aberration. Yet mutations are less errors than they are experiments—variations that introduce novelty into a gene pool. Most mutations are neutral, neither beneficial nor harmful. Some, admittedly, are catastrophic. But occasionally, a mutation confers an advantage: an extra layer of protection against ultraviolet radiation, a heightened sensitivity to sound, or a slight improvement in metabolic efficiency. These rare gifts become the raw material for evolution, passed down through generations and amplified by natural selection.

The interplay between mutation and natural selection produces genetic change, a process so gradual and incremental that it can be difficult to perceive. But evidence of genetic change is etched into the DNA of every living organism. It is visible in the way certain populations of humans have developed resistance to diseases like malaria or how lactose tolerance—a trait absent in most of our ancestors—has become widespread in dairy-consuming cultures.

Evolutionary change is not uniform. It ebbs and flows, accelerating in times of environmental upheaval and slowing during periods of stability. The fossil record offers glimpses into this rhythm. Species emerge, thrive, and disappear, their genetic legacies absorbed into the next iteration of life. These transitions are not arbitrary but guided by the law of evolution, which ensures that the genetic deck is reshuffled continuously, producing new combinations that might succeed where old ones have faltered.

Evolution and Creation Different Questions, One Path

The law of evolution also operates on scales both grand and minute. It governs the sprawling transitions from single-celled organisms to multicellular life, but it also influences the minuscule adaptations within a single species. Take, for instance, the peppered moth in industrial-era England. Before industrialization, light-colored moths blended seamlessly with lichen-covered trees, while darker moths were conspicuous targets for predators. As soot from factories darkened the trees, the balance shifted. Darker moths, once a disadvantage, became the favored form, their survival secured by the changing environment. This shift occurred within a matter of decades—an astonishingly brief moment in evolutionary terms.

In the context of human life, the law of evolution continues to shape our biological and cultural trajectory. Antibiotic resistance in bacteria—a direct consequence of natural selection—is an unsettling reminder of evolution's power. The overuse of antibiotics has created environments where only the most resilient bacteria survive and proliferate. These "superbugs" are not theoretical threats; they are very real, and they demonstrate how human actions can inadvertently accelerate evolutionary processes.

The beauty of the law of evolution lies in its universality. It applies to every organism, from the simplest bacterium to the most complex mammal. It transcends the boundaries of species and kingdoms, weaving all life into a single narrative of change and adaptation. Understanding this law is not merely an academic exercise; it is an invitation to see the world through a lens of interconnectedness and transformation.

By observing the law of evolution, we glimpse the continuity of life—a thread that connects us to the earliest microbes and stretches forward into an uncertain future. It reminds us that change is not only inevitable but essential, a mechanism through which life persists and thrives. To understand the law of evolution is to understand a fundamental truth about existence: that life is, and always has been, in flux.

Evolution and Creation Different Questions, One Path

The Theory of Evolution: From Darwin and Mendel to the Modern Synthetic Theory

A theory, in the realm of science, is not a guess or a hunch. It is a framework—a meticulously crafted map that explains how and why certain phenomena occur. In the case of evolution, theories are the bridges that connect the observable laws of genetic inheritance and natural selection to the mechanisms that produce the rich diversity of life. Theories evolve, much like the organisms they describe, growing more nuanced and comprehensive as new discoveries illuminate previously unseen connections.

The story of evolutionary theory begins with Charles Darwin, but he is far from its sole protagonist. Darwin did not invent the idea of evolution; thinkers before him had grappled with the notion that life changes over time.

Yet, Darwin gave evolution its most compelling narrative: the theory of natural selection. In his seminal work, On the Origin of Species, Darwin proposed that individuals within a species vary in traits, some of which make them better suited to survive and reproduce in their environments. These advantageous traits, passed down through generations, gradually reshape populations, creating new species over vast stretches of time.

Darwin's brilliance lay in his observations—meticulous notes on finches in the Galápagos Islands, fossilized remains of extinct creatures, and the striking adaptations of orchids and insects. He presented natural selection as a slow, relentless sculptor, chipping away at the unfit while refining those that thrive. Yet Darwin, for all his insights, lacked a critical piece of the puzzle: the mechanism of heredity. How were traits passed from one generation to the next? Without an answer, his theory rested on incomplete foundations.

Enter Gregor Mendel, an Austrian monk with a penchant for peas. Working quietly in a monastery garden, Mendel conducted experiments that would revolutionize biology. By crossbreeding pea plants with specific traits—height, color, and seed shape—Mendel uncovered the rules of genetic inheritance. Traits, he found, were passed down in discrete units, later termed genes. Dominant and recessive traits interacted in predictable ways, forming patterns that could be calculated and tested. Mendel's work, published in 1866, went largely unnoticed during his lifetime, overshadowed by Darwin's theory. It wasn't until the early 20th century that Mendel's findings were rediscovered, merging with Darwin's framework to create a more complete picture of evolution.

The marriage of Darwin's natural selection and Mendel's genetics marked the birth of what is now called the Modern Synthesis, or the synthetic theory of evolution. This mid-20th century paradigm integrated insights from various disciplines—paleontology, population genetics, ecology, and embryology—into a unified explanation of how evolution operates. It recognized mutation as the source of genetic variation, natural selection as the filter, and heredity as the conveyor belt through which traits move across generations.

Mutation, once misunderstood as a rare and often harmful aberration, became central to the Modern Synthesis. Scientists realized that mutations introduce the raw material for evolution—a steady trickle of genetic changes that fuel the endless variety of life. Most mutations are neutral, some deleterious, but a select few offer a survival advantage. Over generations, these advantageous mutations accumulate, driving the gradual evolution of populations.

Population genetics brought a mathematical precision to evolutionary theory. Scientists like Ronald Fisher and J.B.S. Haldane quantified how genetic traits spread through populations, linking Darwinian selection with Mendelian inheritance. Their work demonstrated that evolution is not a linear march toward perfection but a dynamic process shaped by chance, adaptation, and environmental pressures. The Modern Synthesis also accounted for the fossil record's apparent gaps, showing that evolution is neither uniformly slow nor episodically abrupt but a complex interplay of gradual change and punctuated equilibrium.

Despite its elegance, the Modern Synthesis is not the final word on evolution. In recent decades, new discoveries have expanded and challenged its framework. Epigenetics, for instance, has revealed that environmental factors can influence gene expression without altering DNA sequences—a wrinkle Darwin and Mendel could scarcely have imagined. Horizontal gene transfer, the exchange of genetic material between unrelated species, has upended traditional views of inheritance. And advances in molecular biology have uncovered layers of complexity within cells, from the intricate dance of proteins to the self-editing capabilities of RNA.

These discoveries do not undermine the theory of evolution; they enrich it. Evolution is not a rigid doctrine but a living, breathing discipline, open to revision and refinement. It is a testament to the power of science—its willingness to question, adapt, and grow. Darwin's finches and Mendel's peas are joined by CRISPR gene-editing, genome sequencing, and artificial intelligence in an ever-expanding narrative that seeks to understand life's complexity.

The theory of evolution, in all its iterations, is more than a scientific achievement. It is a lens through which we view the world, a story that connects every organism in

Evolution and Creation Different Questions, One Path

a web of shared ancestry. From the ancient stromatolites that first breathed life into Earth's barren oceans to the intricate neural networks of the human brain, evolution offers an explanation for the improbable miracle of existence. It invites us to marvel at the connections that bind us to every living thing, to see ourselves not as isolated beings but as threads in a vast, interwoven tapestry.

As the Modern Synthesis evolves into its next iteration, it reminds us that science, like life, thrives on change. The questions we ask today may yield answers that reshape our understanding tomorrow. Yet, amidst this flux, the theory of evolution stands as a beacon—a reminder that the journey of discovery is as profound as the destination.

Kendir Ramiz

The Illusion of Conflict: Why Science and Faith Are Not at Odds

Conflict thrives on false dichotomies. Nowhere is this more evident than in the perennial debate between science and faith. The two are often portrayed as antagonistic forces, locked in an eternal struggle for dominance over humanity's understanding of the universe. Yet, this supposed clash is less a battle of substance and more a mirage born from misunderstanding. Science and faith, rather than existing in opposition, are distinct lenses through which we attempt to unravel the mysteries of existence.

Science, at its core, seeks to answer the "how" of the universe. It is a method, a systematic inquiry designed to uncover patterns, test hypotheses, and refine knowledge. Its power lies in its humility—it does not

Evolution and Creation Different Questions, One Path

claim absolute truth but instead invites constant questioning and revision. The laws of physics, the mechanisms of evolution, the origins of the cosmos—these are the domains of science, observable and measurable phenomena that reveal the workings of the natural world.

Faith, on the other hand, addresses the "why." It delves into purpose, morality, and meaning, offering frameworks for understanding existence beyond what can be quantified or dissected. Faith is deeply personal, rooted in tradition, culture, and individual experience. For billions, it provides not just answers but a sense of connection to something greater than themselves—a divine presence, an eternal truth, or an overarching order.

The perceived conflict between science and faith often arises from conflating these distinct domains. When science attempts to answer existential questions of purpose, or when faith ventures into empirical claims about the natural world, tensions inevitably surface. A scientist insisting that the universe is devoid of meaning because it can be explained through natural laws is no less misplaced than a theologian rejecting evolutionary biology on the grounds of scriptural interpretation. Both miss the essence of their respective disciplines.

Consider evolution itself, a frequent flashpoint in this supposed conflict. Some see it as an affront to creation, a denial of divine agency in the unfolding of life. Yet, to view evolution and creation as mutually exclusive is to misunderstand both. If one believes in a creator, could not evolution be the mechanism through which creation unfolds? The genetic code, with its ability to mutate and adapt, could just as easily be seen as a testament to divine ingenuity as to natural processes. Evolution need not diminish the concept of a creator; it may, in fact, illuminate it.

History is replete with figures who embraced both science and faith without contradiction. Isaac Newton, often regarded as the father of modern science, saw his groundbreaking work on gravity and motion as a means of understanding the divine order of the cosmos. Gregor Mendel, whose experiments with pea plants laid the foundation for modern genetics, was a monk whose faith informed his curiosity about the natural world. More recently, Francis Collins, the scientist who led the Human Genome Project, described his work as "an opportunity to glimpse at the instructions set down by the Creator."

These examples remind us that science and faith are not rival factions but complementary pursuits. Science provides tools for understanding the mechanics of existence, while faith grapples with its deeper meanings. Together, they form a holistic framework, offering insights that neither could achieve alone.

The illusion of conflict is further perpetuated by societal narratives that thrive on polarization. In the age of sound bites and clickbait, nuance is often sacrificed for the allure of controversy. A headline proclaiming a "war"

Evolution and Creation Different Questions, One Path

between science and religion is more likely to capture attention than one exploring their interplay. This distortion fosters division, creating the false impression that one must choose between the two.

Yet, when we step away from these divisive narratives, a more harmonious picture emerges. Science and faith, when allowed to operate within their respective spheres, enhance rather than undermine one another. Science grounds us in the tangible, revealing the intricate mechanisms that govern the natural world. Faith lifts us beyond the tangible, inviting contemplation of purpose and morality. Together, they address both the how and the why, forming a bridge between understanding and meaning.

This interplay is perhaps most evident in the study of the cosmos. The Big Bang theory, which describes the universe's origins, is a triumph of scientific discovery. It traces the expansion of space and time, the birth of matter and energy. Yet, the question of what preceded the Big Bang remains unanswered by science. For some, this is where faith steps in—not as an adversary to scientific inquiry but as a complement, addressing the profound mystery of existence itself.

The false dichotomy between science and faith not only undermines both but also limits our capacity for wonder. To reduce science to mere mechanics or faith to rigid dogma is to strip them of their transformative power. Science invites awe at the natural world's complexity, from the elegance of DNA to the vastness of the cosmos. Faith nurtures a sense of belonging and purpose, a recognition that we are part of something greater.

By embracing the strengths of both, we enrich our understanding of the universe and our place within it. Science and faith are not adversaries but partners in the human quest for knowledge and meaning. They remind us that the search for truth is not confined to a single path but is an expansive journey, weaving together reason and reverence, curiosity and contemplation.

The illusion of conflict dissolves when we recognize this unity. Science and faith, far from being mutually exclusive, are two sides of the same coin—a testament to humanity's insatiable desire to comprehend the world and our role within it. Together, they challenge us to see not just with our eyes but with our minds and hearts, to seek not just understanding but wisdom.

Evolution and Creation Different Questions, One Path

Understanding Humanity Through Evolution: Vestiges and Traces in the Human Body

The human body is a paradox. It is a marvel of complexity, capable of extraordinary feats, yet it bears within itself the unmistakable marks of its evolutionary past. To truly understand what it means to be human, we must look not only at our achievements and aspirations but also at the traces left behind by the long journey of evolution. These vestiges and remnants—what some might call "leftovers" of our evolutionary ancestors—are silent witnesses to the millions of years of adaptation that shaped us into who we are today.

Evolution is a story written in anatomy. Every bone, muscle, and organ in the human body tells a tale of survival, adaptation, and sometimes obsolescence. Consider, for example, the palmaris longus, a tendon in the wrist that serves little to no purpose in modern humans. In some individuals, it is completely absent. Yet in our primate relatives, this tendon plays a crucial role, aiding in their ability to swing and climb. Its presence in some humans and absence in others is not an error but evidence of evolution in progress—a once vital structure gradually fading into irrelevance as our lifestyle diverged from that of our arboreal ancestors.

Similarly, the appendix, often dismissed as a vestigial organ, has its origins in a far more active past. In herbivorous ancestors, the appendix was part of a larger cecum, essential for digesting cellulose-rich plant material. As our diet shifted and our reliance on raw vegetation diminished, the appendix shrank, losing its original purpose. Today, it is a small, finger-like pouch that occasionally causes trouble, yet some researchers believe it may still serve a secondary role in housing beneficial gut bacteria. Its transformation is a microcosm of evolution's balancing act: preserving what is useful, discarding what is not, and sometimes repurposing the remnants.

The traces of our evolutionary heritage are not confined to organs. They are etched into the very structure of our skeletons. The coccyx, or tailbone, is a particularly striking example. This small cluster of fused vertebrae is all that remains of the tails our distant mammalian ancestors once used for balance and communication. While it no longer serves its original purpose, the coccyx occasionally reminds us of its past glory through rare instances of human babies born with small, functional

Evolution and Creation   Different Questions, One Path

tails—a genetic echo from an earlier chapter of life on Earth.

Even the muscles we carry reflect a bygone era. The arrector pili, tiny muscles attached to hair follicles, are responsible for the phenomenon of goosebumps. In our evolutionary past, these muscles served a practical purpose: they would contract to raise the hair on our bodies, creating a layer of insulation in the cold or making us appear larger to predators when threatened. Today, in our relatively hairless state, goosebumps are little more than a fleeting reminder of the days when our ancestors relied on fur for survival.

These evolutionary relics are not merely curiosities; they are keys to understanding our relationship with the natural world. They remind us that we are not static beings, but part of a continuum—a lineage stretching back billions of years. Our bodies are not perfect machines designed from scratch but patchworks of adaptations, each one shaped by the pressures and constraints of its time.

Consider the inefficiencies of the human spine. Its S-shaped curve, so prone to pain and injury, is a compromise—a solution to the problem of walking upright while retaining the basic structure of a quadrupedal ancestor. Similarly, the human knee, though capable of remarkable feats of endurance, is a fragile joint, vulnerable to wear and injury because it evolved to support a body that was once more horizontal than vertical. These imperfections are not flaws in design but artifacts of evolution's incremental nature. Evolution does not build from scratch; it modifies what already exists, often repurposing structures in ways that are functional but far from ideal.

Even our senses carry echoes of the past. The plica semilunaris, a small fold of tissue in the corner of the human eye, is a remnant of a nictitating membrane—a third eyelid that once swept horizontally across the eyes of our ancestors, protecting and moistening them. While it serves no practical purpose today, its persistence in our anatomy is a testament to the evolutionary principle that change is slow and often incomplete.

These vestiges are not just biological trivia; they are profound reminders of our place in the grand narrative of life. They anchor us to a history that predates humanity, connecting us to the countless species that came before. Each vestigial structure, each evolutionary quirk, tells a story of resilience and adaptation, of life finding ways to persist and thrive in the face of changing environments.

Yet, these traces also raise questions. Why do some traits persist while others vanish? What determines whether a structure is retained, repurposed, or discarded? Evolution offers no definitive answers, only the ongoing dance of genetic variation and natural selection. What is clear, however, is that our evolutionary journey is far from over. As environments change, as lifestyles shift, and as technology reshapes the pressures we face, the human body will continue to evolve, its anatomy reflecting the challenges and opportunities of the future.

To study these evolutionary remnants is to engage with the deepest roots of what it means to be human. They are not relics to be dismissed but clues to our origins and signposts to our potential. In every vestigial tendon, muscle, and organ lies a reminder of the profound continuity of life—a testament to the power of change and the resilience of existence. Through these traces, we come to see ourselves not as isolated beings but as part of a vast and interconnected web, stretching from the first spark of life to the boundless possibilities of what lies ahead.

Kendir Ramiz

The First Thoughts on Evolution: The Legacy of the Upanishads, Anaximander, and Empedocles

The idea of evolution did not spring fully formed from the mind of Charles Darwin. Long before the advent of modern science, human beings grappled with the question of how life changes, adapts, and transforms. Across cultures and epochs, thinkers and philosophers sought to make sense of the natural world and its seemingly endless capacity for renewal and variation. These early musings on evolution may lack the precision of contemporary science, but they reveal a profound curiosity about the mechanisms underlying life—a curiosity that echoes across millennia.

In the ancient texts of India, particularly the Upanishads, we find one of the earliest philosophical explorations of change and transformation. These texts, written thousands of years ago, delve into the nature of existence, the cycles of creation and destruction, and the interconnectedness of all things. The Upanishads speak of the universe as a dynamic, ever-changing

Evolution and Creation Different Questions, One Path

entity, where life emerges, dissolves, and re-emerges in an eternal cycle. This perspective is not a scientific hypothesis but a metaphysical framework—a way of understanding the fluid and impermanent nature of reality. The idea that all life is part of a larger, interconnected whole resonates with the later insights of evolutionary biology, even if the language and approach differ.

While the Upanishads offered a philosophical lens, the ancient Greeks sought more concrete explanations for life's diversity. Among the pre-Socratic philosophers, Anaximander of Miletus stands out as one of the earliest thinkers to propose a naturalistic account of the origins of life. Writing in the 6th century BCE, Anaximander suggested that life began in water and that early creatures gradually adapted to their environments. He hypothesized that humans may have evolved from aquatic animals, reasoning that our survival as infants depends on extended parental care, unlike most other animals. Though primitive by modern standards, Anaximander's ideas represent a remarkable attempt to understand life as a process shaped by natural forces rather than divine intervention.

Building on this tradition, Empedocles of Acragas in the 5th century BCE introduced a concept that closely resembles natural selection. Empedocles envisioned the world as a chaotic amalgamation of elements—earth, air, fire, and water—held together by the opposing forces of Love and Strife. In this ever-changing environment, creatures emerged through a process of trial and error. Empedocles proposed that only those forms capable of survival persisted, while others—those with mismatched or dysfunctional parts—faded into oblivion. This notion, though metaphorical, bears a striking resemblance to Darwin's later principle of natural selection, where the environment determines which traits and species endure.

These early ideas are not isolated fragments but part of a broader tapestry of human thought. They reveal an intuitive understanding that life is dynamic, shaped by forces larger than any single individual or species. What is striking is how these thinkers, separated by geography and culture, arrived at ideas that foreshadowed modern evolutionary theory. They lacked the tools of genetics, paleontology, and molecular biology, yet their observations and reflections captured essential truths about the nature of life.

The significance of these early thinkers lies not just in their specific ideas but in their willingness to question, to observe, and to hypothesize. They represent the birth of a mindset that values inquiry over dogma, a mindset that paved the way for the scientific revolution centuries later. In their writings, we see the seeds of a tradition that would come to define science itself: the relentless pursuit of understanding through observation, reasoning, and experimentation.

It is easy to dismiss these early musings as naive or speculative, but to do so would be to overlook their profound contribution to the intellectual heritage of

Evolution and Creation Different Questions, One Path

humanity. The Upanishads, Anaximander, and Empedocles remind us that the journey to understanding evolution did not begin in a laboratory but in the human capacity for wonder. They teach us that science is not merely a collection of facts but a way of seeing the world—one that begins with the simple yet profound act of asking, "Why?"

This legacy continues to shape how we approach the mysteries of life. Modern evolutionary biology, with its rigorous methods and detailed analyses, builds on the foundations laid by these early thinkers. The questions they posed—about the origins of life, the mechanisms of change, and the interconnectedness of all things—remain central to the scientific endeavor. As we delve deeper into the complexities of evolution, we carry with us the insights of those who first dared to imagine a world shaped by transformation and adaptation.

By tracing the lineage of evolutionary thought, we see that science is not a series of isolated discoveries but a continuous dialogue across time and culture. The Upanishads, Anaximander, and Empedocles may not have had the tools to test their ideas, but their willingness to explore the unknown laid the groundwork for the discoveries that followed. They remind us that the search for knowledge is not confined to any one era or discipline but is a shared human endeavor—a testament to our innate curiosity and our enduring quest to understand the world around us.

Kendir Ramiz

Al-Jahiz and the Discovery of Adaptation in Nature

When discussing the history of evolutionary thought, much attention is paid to the ancient philosophers of Greece and the scientific breakthroughs of Renaissance Europe. Yet, there is a critical chapter in this story that is often overlooked: the intellectual flowering of the Islamic Golden Age. Between the 8th and 13th centuries, while much of Europe languished in what are often referred to as the "Dark Ages," scholars in the Islamic world were advancing knowledge across disciplines—mathematics, astronomy, medicine, and yes, even early biology. Central to this intellectual movement was a remarkable figure whose insights predated Darwin by over a millennium: Al-Jahiz.

Evolution and Creation Different Questions, One Path

Born in Basra around 776 CE, Abu Uthman Amr ibn Bahr Al-Jahiz was a polymath—a writer, philosopher, and naturalist whose curiosity about the world knew no bounds. His most famous work, Kitab al-Hayawan (The Book of Animals), spans seven volumes and is a rich tapestry of observations on zoology, ecology, and the interconnectedness of life. At first glance, The Book of Animals might seem like a medieval bestiary, cataloging the habits and characteristics of various creatures. But beneath its surface lies a proto-evolutionary framework, one that anticipates key concepts of adaptation and survival in a changing environment.

Al-Jahiz's most profound contribution lies in his understanding of what we now call ecological relationships. He observed that living organisms do not exist in isolation but are part of complex systems where they interact with one another and with their surroundings. Predators hunt prey, prey adapt to avoid predators, and competition for resources drives species to innovate or perish. Al-Jahiz described this interplay in vivid detail, writing of how "strong animals prey on the weak, and large animals devour the small." He noted that these struggles for survival create a balance in nature, where every organism has a role to play in the larger ecosystem.

What makes Al-Jahiz's insights particularly striking is his recognition of adaptation as a dynamic process. He observed that animals develop traits that help them survive in specific environments—a rudimentary understanding of natural selection. For example, he wrote about how certain birds adapted their beaks and behaviors to better access food, a concept that would later become a cornerstone of Darwin's work. Al-Jahiz's writings reveal a deep awareness of the fluidity of life, where change is not a rare anomaly but a constant force shaping the natural world.

Al-Jahiz also explored the idea of inherited traits, though his framework was more philosophical than scientific. He speculated that the characteristics of animals could be passed down through generations, influenced by their interactions with the environment. While he lacked the tools and knowledge to delve into genetics—a field that would not emerge until Mendel's experiments centuries later—his intuition laid the groundwork for thinking about heredity as a mechanism for evolutionary change.

Beyond his scientific observations, Al-Jahiz's work reflects a broader worldview rooted in the Islamic intellectual tradition. In his writings, he frequently linked his observations of nature to a sense of divine order. For Al-Jahiz, the complexity and harmony of the natural world were signs of a higher intelligence—a creator who designed the universe with purpose and balance. This perspective did not conflict with his scientific inquiries but complemented them, illustrating how faith and reason coexisted in the Islamic Golden Age.

The broader context of Al-Jahiz's work is equally important. He was part of a vibrant intellectual culture centered in cities like Baghdad and Basra, where

Evolution and Creation Different Questions, One Path

scholars translated and preserved the works of ancient Greek philosophers while also producing original contributions to science and philosophy. This era saw the establishment of institutions like the House of Wisdom, where thinkers from diverse backgrounds collaborated to expand human knowledge. Al-Jahiz's work is a testament to this spirit of inquiry, blending observation, analysis, and philosophical reflection in a way that feels remarkably modern.

Yet, despite his groundbreaking insights, Al-Jahiz's contributions are often overshadowed in Western narratives of evolutionary thought. His name rarely appears in textbooks or discussions of pre-Darwinian ideas, a reflection not of his significance but of the biases that shape how history is told. To fully appreciate the development of evolutionary theory, we must recognize figures like Al-Jahiz, whose ideas were not isolated but part of a broader continuum of human curiosity about the natural world.

The legacy of Al-Jahiz extends beyond his specific observations. His work exemplifies a way of thinking that is both holistic and dynamic, emphasizing the interconnectedness of life and the processes that drive its diversity. It challenges us to view adaptation not as a static phenomenon but as an ongoing dialogue between organisms and their environments. And it reminds us that the history of science is not the story of any one culture or era but a collective endeavor, enriched by the contributions of thinkers from every corner of the world.

Kendir Ramiz

By studying the work of Al-Jahiz and his contemporaries, we gain not only a deeper understanding of the roots of evolutionary thought but also a renewed appreciation for the intellectual heritage that unites humanity. His insights, written in the margins of history, continue to resonate today, offering a perspective on life that is both timeless and urgently relevant. Evolution, as Al-Jahiz saw it, is not merely a process but a reflection of life's endless capacity for change—a reminder that the natural world, like knowledge itself, is always in motion.

The Renaissance and the Contributions of Erasmus Darwin

The Renaissance, often hailed as the great cultural and intellectual revival of Europe, was more than just an era of rediscovered art and architecture. It marked the rekindling of curiosity—a curiosity that had lain dormant during centuries of dogma-driven stagnation. As ancient texts were unearthed and translated, a new generation of thinkers began to question the world around them, breaking free from the constraints of rigid orthodoxy. This intellectual upheaval laid the groundwork for scientific revolutions to come, including the evolution of evolutionary thought itself.

One of the most remarkable aspects of the Renaissance was its ability to bridge the old and the new. Thinkers such as Leonardo da Vinci and Galileo Galilei delved into anatomy, astronomy, and mechanics, guided by the rediscovered works of ancient Greek and Roman scholars. The Renaissance was not merely an age of revival; it was an age of synthesis. By combining the empirical approaches of antiquity with emerging scientific methods, Renaissance scholars created fertile ground for future breakthroughs.

This renewal of inquiry, however, did not stop with the Renaissance. Its spirit carried forward into the Enlightenment and beyond, culminating in figures like Erasmus Darwin, the grandfather of Charles Darwin. Erasmus Darwin lived and wrote in 18th-century England, a time when intellectual curiosity was blossoming across Europe. Although he is often overshadowed by his grandson, Erasmus was a pivotal figure in his own right—a poet, physician, inventor, and naturalist who dared to imagine a world governed by the principles of change and adaptation.

Erasmus Darwin's most notable contribution to evolutionary thought came in the form of his poetry, particularly in his work Zoonomia (1794-1796). While poetry might seem an unlikely medium for scientific exploration, Erasmus Darwin used it to communicate complex ideas about biology, life, and transformation in ways that were both accessible and thought-provoking. In Zoonomia, he proposed that all living things share a common ancestry and that they change over time in response to their environments. This idea was revolutionary in its implications, challenging the prevailing belief in the immutability of species.

Unlike his grandson Charles, who would later base his theory on meticulous empirical evidence, Erasmus Darwin relied on intuition and observation. He noticed patterns in the natural world that suggested a deep interconnectedness among living beings. For instance, he observed the anatomical similarities between different species and speculated that these commonalities pointed to a shared origin. He wrote, "Would it be too bold to imagine that all warm-blooded animals have arisen from one living filament?" This single question encapsulates the essence of his thinking—a bold and imaginative leap that prefigured the scientific rigor of the 19th century.

Erasmus Darwin also explored the concept of adaptation, though his understanding was more poetic than scientific. He speculated that organisms respond to their environments in ways that shape their development, an idea that hinted at natural selection without fully articulating its mechanisms. He envisioned evolution as a creative force, driven by the interactions between organisms and their surroundings. In one of his verses, he described nature as an artist, sculpting life through countless experiments and variations—a metaphor that resonates even in modern evolutionary biology.

What makes Erasmus Darwin's contributions particularly significant is their timing. He lived in an era when the very idea of species change was heretical, a direct challenge to the prevailing religious worldview. Yet he approached these questions not with defiance but with curiosity, blending science, philosophy, and art into a coherent vision of life's complexity. His willingness to entertain radical ideas, even without definitive proof, exemplifies the exploratory spirit of his age.

The legacy of Erasmus Darwin is not merely one of scientific speculation but of intellectual daring. He paved the way for future thinkers by normalizing the idea that life is dynamic and ever-changing. His writings inspired not only his grandson Charles but also a generation of naturalists and philosophers who would expand and refine his ideas. The concept of common ancestry, which he hinted at in Zoonomia, would later become a cornerstone of modern biology.

The Renaissance and the Enlightenment were not just periods of rediscovery; they were periods of reimagination. Erasmus Darwin embodied this spirit, looking at the world not as a fixed creation but as a canvas of infinite possibility. He reminds us that progress is often born from the courage to ask unconventional questions, even when the answers are uncertain.

The story of evolutionary thought is incomplete without acknowledging Erasmus Darwin's poetic speculations and the broader intellectual currents of his time. Together, they demonstrate that the journey to understanding evolution was not a linear path but a winding road, shaped by the interplay of science, philosophy, and creativity. Through the works of thinkers like Erasmus, we see that evolution is not just a scientific theory but a profound narrative—one that connects us to the past and invites us to imagine the future.

Evolution and Creation Different Questions, One Path

Rewriting the History of Evolution Through Archaeological Discoveries

The ground beneath our feet holds secrets that stretch back billions of years. Fossils—those remnants of ancient life—are not just curiosities to be admired in museums. They are the pages of Earth's biography, chronicling a story of transformation and resilience. Each fossil represents a moment frozen in time, a fragment of a puzzle that, when pieced together, reveals the grand narrative of evolution.

The discovery of fossils predates modern science. For centuries, they were misunderstood, often dismissed as mere stones or seen as evidence of mythical creatures. In medieval Europe, fossilized remains of large animals were frequently attributed to dragons or giants, their true origins obscured by a lack of understanding. But as the Renaissance ignited a renewed interest in the natural world, fossils began to be recognized as something far more profound: the preserved remains of creatures that once roamed the Earth.

The study of fossils—what we now call paleontology—emerged as a discipline in the 18th and 19th centuries, propelled by groundbreaking discoveries that challenged prevailing notions about the world. Fossils provided tangible evidence that life on Earth was not static but dynamic, subject to processes of change and extinction. This realization was revolutionary. It suggested that the world was far older than previously thought and that its history was one of continuous transformation.

One of the pivotal figures in this narrative was Georges Cuvier, a French naturalist who pioneered the field of comparative anatomy. Cuvier demonstrated that fossils belonged to species no longer found on Earth, providing the first concrete evidence of extinction. His meticulous studies of fossilized bones and teeth allowed him to reconstruct the anatomy of ancient creatures, painting vivid pictures of a world long past. Although Cuvier rejected the idea of evolution, his work laid the groundwork for future scientists to explore the mechanisms behind the changes fossils so clearly documented.

As paleontology advanced, fossils became more than just isolated relics. They were recognized as part of a vast, interconnected record of life, spanning millions of years. The discovery of transitional fossils—those that display traits bridging the gap between major groups of organisms—was particularly transformative. These fossils, such as Archaeopteryx, which exhibits both

Evolution and Creation Different Questions, One Path

reptilian and avian characteristics, offered compelling evidence for the gradual evolution of species. They were the missing links that Darwin's theory of natural selection had predicted but which had not yet been found when On the Origin of Species was published.

Fossils also revolutionized our understanding of the human story. The unearthing of early hominin remains, such as Lucy, the famous Australopithecus afarensis, revealed a lineage of bipedal primates that connected modern humans to their distant ancestors. Lucy's discovery in 1974 was a watershed moment, providing a glimpse into a period of evolutionary history when walking upright began to shape the anatomy and behavior of our ancestors. Her skeleton, remarkably preserved, spoke volumes about the adaptations that set the stage for the emergence of Homo sapiens.

The fossil record, however, is far from complete. It is a patchwork, with gaps that reflect the rarity of fossilization—a process dependent on specific conditions that allow organic material to be preserved over vast spans of time. Despite these gaps, the fossils we have unearthed tell a remarkably coherent story. They show patterns of divergence and convergence, of extinctions and survivals, of adaptation and innovation. They reveal that evolution is not a straight line but a branching tree, with countless paths explored and abandoned over the course of life's history.

Archaeological discoveries continue to reshape our understanding of evolution. Advances in technology, such as CT scanning and isotope analysis, allow scientists to extract new information from old fossils. These tools can reveal details about an organism's diet, habitat, and even its behavior, adding depth and nuance to the fossil record. Recent discoveries, such as feathered dinosaurs in China or ancient footprints preserved in volcanic ash, challenge and refine our understanding of how life evolved and adapted to changing environments.

But fossils are not just scientific artifacts; they are profound reminders of the passage of time and the impermanence of life. They show us that extinction is not the exception but the rule—that most species that have ever lived are now gone. This perspective can be humbling, even unsettling. Yet it also underscores the resilience of life, its ability to adapt and diversify in the face of immense challenges.

The story of fossils is not just a story of the past; it is a story that continues to unfold. Each new discovery adds a chapter, offering fresh insights and raising new questions. Fossils challenge us to think about our place in the grand sweep of Earth's history. They remind us that humanity is but a recent addition to the evolutionary tapestry, a single thread in a pattern that stretches back billions of years.

Through fossils, we come to see evolution not as an abstract theory but as a tangible reality, etched into the very fabric of the Earth. They connect us to the ancient past, revealing a world both alien and familiar. They

Evolution and Creation Different Questions, One Path

invite us to marvel at the complexity and beauty of life, to reflect on the forces that have shaped it, and to consider what the future may hold. In the story of fossils, we find not just the origins of species but the origins of wonder itself—a testament to the enduring power of discovery and the endless curiosity that drives us to uncover the secrets of the Earth.

Kendir Ramiz

The Roots of the Evolutionary Tree

Life on Earth is a masterpiece of interconnectedness. Every species, from the simplest bacterium to the most complex mammal, exists as part of an intricate web—a vast evolutionary tree whose roots stretch back billions of years. This tree, often depicted as a branching diagram in biology textbooks, is more than a scientific abstraction. It is a map of life itself, tracing the origins of every organism and showing how each is connected through shared ancestry. To explore the roots of this tree is to delve into the very essence of existence, a journey that reveals both the unity and diversity of life.

The concept of an evolutionary tree is not new. Its earliest iterations can be found in the musings of ancient philosophers and naturalists who observed patterns in the natural world. Yet, it wasn't until the 19th century,

Evolution and Creation Different Questions, One Path

with Charles Darwin's publication of On the Origin of Species, that the metaphor of the tree truly took shape. Darwin himself sketched a rudimentary tree in his notebooks, a simple diagram with branching lines that symbolized the divergence of species over time. He wrote, "I think," above the sketch—a modest prelude to a revolutionary idea. Life, Darwin proposed, is not a series of isolated creations but a single, branching lineage.

At the base of this tree lie the roots—the origin of life itself. Though the precise details of life's beginnings remain shrouded in mystery, scientists have pieced together a plausible narrative. It begins in Earth's primordial past, some 3.8 billion years ago, when the planet's oceans teemed with organic molecules. These molecules, energized by heat and light, coalesced into self-replicating structures—the first forms of life. From these humble beginnings arose prokaryotes, single-celled organisms that dominated the planet for billions of years. They were the architects of life's early chapters, laying the foundation for everything that would follow.

As life evolved, it diversified, branching into new forms and filling ecological niches. The invention of photosynthesis, for example, transformed the planet, allowing organisms to harness sunlight and produce oxygen as a byproduct. This oxygenation event, known as the Great Oxidation Event, reshaped Earth's atmosphere and paved the way for more complex life forms. Eukaryotes—organisms with complex cells containing a nucleus—emerged, leading to the rise of multicellular life. Each new adaptation was like a branch on the evolutionary tree, expanding its reach and complexity.

What makes this tree remarkable is its universality. It includes every living organism, connecting humans to plants, fungi, and even microscopic bacteria. At the molecular level, this connection is undeniable. All life shares a common genetic language, encoded in DNA. The same four nucleotides—adenine, thymine, cytosine, and guanine—compose the genetic instructions for everything from a blade of grass to a blue whale. This shared code is a testament to life's unity, a reminder that every species is part of the same ancient lineage.

Yet, the tree of life is not static. It is constantly growing and reshaping itself, responding to the pressures of natural selection, genetic drift, and environmental change. Some branches flourish, giving rise to an explosion of diversity, while others wither and fade, leaving only traces in the fossil record. Mass extinctions—such as the Permian-Triassic extinction, which wiped out nearly 90% of species—serve as stark reminders of life's fragility. But they also highlight its resilience. After each extinction event, the tree of life rebounded, producing new forms that filled the void left by those that perished.

The interconnectedness of life extends beyond biology. It is woven into the fabric of ecosystems, where species depend on one another in intricate networks. A single

Evolution and Creation Different Questions, One Path

tree in a rainforest, for instance, may host hundreds of species—birds, insects, fungi, and microorganisms—all interacting in a delicate balance. These relationships are not merely incidental; they are the product of millions of years of co-evolution, a process that binds species together in mutual dependence. The loss of one species can ripple through the entire ecosystem, disrupting the balance and threatening the survival of others.

The evolutionary tree also offers profound insights into humanity's place in the natural world. Humans are not perched at the top of this tree, as earlier models once suggested, but are part of a branch that includes other primates, such as chimpanzees and gorillas. This branch, in turn, connects to older branches that trace back to a common ancestor shared with all mammals, reptiles, and amphibians. Far from diminishing our uniqueness, this perspective enriches it. It shows that our story is deeply intertwined with the stories of countless other species—a shared history of adaptation and survival.

Understanding the roots of the evolutionary tree challenges us to think differently about life. It invites us to see not just the differences between species but the common threads that unite them. It reminds us that every living thing, from the smallest microbe to the largest elephant, is part of a single, ongoing experiment—a testament to life's capacity for change and innovation.

As we explore this tree, we also confront the responsibility that comes with knowledge. Humanity, with its unprecedented ability to shape the environment, holds the power to protect or destroy the diversity of life. The choices we make—how we use resources, how we interact with ecosystems, and how we view our relationship with nature—will determine the future of the evolutionary tree. Will it continue to branch and grow, or will it bear the scars of human impact?

The roots of the evolutionary tree are more than a scientific concept; they are a reminder of life's interconnectedness and its infinite potential for renewal. They challenge us to look beyond the surface, to see the invisible threads that bind us to every other living thing. In this interconnected tapestry, we find not just the origins of life but the essence of what it means to belong—to be a part of something vast, enduring, and extraordinary.

Evolution and Creation Different Questions, One Path

The Journey from Orrorin Tugenensis to Homo habilis

To understand humanity's origins, we must venture back millions of years, to a time when the African savannas teemed with life and the ancestors of modern humans were little more than another species navigating the harsh realities of survival. The story of humanity begins not with Homo sapiens but with a lineage that diverged from the common ancestor we share with chimpanzees. It is a story marked by incremental changes—a journey of adaptation, resilience, and discovery. Among the earliest known figures in this epic tale is Orrorin tugenensis, whose fossilized remains whisper of a pivotal moment in evolution: the shift to walking upright.

Discovered in Kenya's Tugen Hills in 2000, Orrorin tugenensis lived approximately six million years ago, at a time when Earth was undergoing significant environmental changes. Fossil evidence suggests that Orrorin walked on two legs, a trait that distinguishes hominins from other primates. This adaptation, known as bipedalism, was a radical departure from the arboreal lifestyle of earlier ancestors. Yet, Orrorin was not exclusively terrestrial. Its skeletal features indicate that it retained the ability to climb trees with ease, blending old and new behaviors in a way that hints at the transitional nature of evolution.

Why did bipedalism emerge? The answer likely lies in the changing landscapes of East Africa. As forests receded and open grasslands expanded, hominins faced new challenges and opportunities. Walking on two legs offered several advantages: it allowed for better visibility over tall grass, freed the hands for carrying tools and food, and reduced energy expenditure over long distances. While Orrorin's bipedalism was not as advanced as that of later species, it represented the first tentative steps toward a defining characteristic of humanity.

The journey from Orrorin to more advanced hominins spanned millions of years, during which evolutionary pressures continued to shape anatomy and behavior. One of the most significant milestones in this journey is represented by Australopithecus afarensis, the species to which the famous fossil "Lucy" belongs. Lucy's discovery in 1974 revolutionized our understanding of early hominins. At approximately 3.2 million years old, her skeleton revealed a creature that walked upright but still possessed features suited for climbing. Lucy's dual adaptation highlights the gradual nature of evolution, where traits are refined over generations rather than appearing abruptly.

As hominins adapted to their changing environments, another transformative shift began to take shape: the expansion of brain size. Larger brains allowed for greater cognitive abilities, but they also came with costs, requiring more energy and complicating childbirth. This trade-off marked the emergence of a new phase in human evolution, one characterized by increased problem-solving skills, social complexity, and tool use. These changes set the stage for the appearance of Homo habilis, often referred to as "the handy man."

Living around 2.4 to 1.4 million years ago, Homo habilis represents a significant departure from its predecessors. The species is notable for its association with stone tools, which provide a window into the cognitive capabilities of early humans. Tools found alongside Homo habilis fossils are simple yet effective—sharp-edged flakes and cores used for cutting meat, scraping hides, and breaking bones to access marrow. These tools marked the beginning of a technological tradition that would culminate in the sophisticated innovations of Homo sapiens.

Homo habilis also exhibited a marked increase in brain size compared to earlier hominins. With a cranial capacity ranging from 510 to 600 cubic centimeters, its brain was significantly larger than that of Australopithecus. This expansion likely facilitated more complex social behaviors and improved communication, laying the groundwork for the development of language. While Homo habilis was not the direct ancestor of modern humans, it occupied a critical position in the evolutionary lineage, bridging the gap between more primitive hominins and the genus Homo.

The journey from Orrorin tugenensis to Homo habilis is not a straight line but a branching path, with multiple species coexisting and experimenting with different adaptations. Fossil discoveries suggest that evolution was not a singular progression but a tapestry of parallel experiments in survival. Some branches, like that of Orrorin, faded into extinction, while others, like Homo habilis, contributed to the mosaic of traits that define humanity.

This journey also reveals the importance of environment in shaping evolution. The fluctuating climates of East Africa created a landscape of uncertainty, forcing hominins to adapt or perish. Those that succeeded did so not because they were inherently superior but because their traits aligned with the demands of their surroundings. Evolution is not a story of progress but of persistence—a relentless process of trial and error that rewards adaptability over perfection.

The first footsteps of our ancestors echo through time, reminding us that the story of humanity began long before the emergence of Homo sapiens. It is a story written in fossils and artifacts, in the shape of a pelvis or the edge of a stone tool. These early chapters of human evolution are not just tales of anatomical change but of ingenuity, resilience, and the unyielding drive to survive. As we trace this journey, we are reminded that our origins are deeply rooted in the natural world, a testament to the power of adaptation and the interconnectedness of life.

Evolution and Creation Different Questions, One Path

The First Steps of Technological Revolution

Long before the printing press, the wheel, or the Internet, humanity's greatest technological breakthroughs were etched not in blueprints but in stone and flame. The creation of stone tools and the mastery of fire marked pivotal moments in the evolutionary journey of our ancestors. These innovations were not merely about survival; they were catalysts for profound changes in anatomy, behavior, and society. The story of stone tools and fire is the story of humanity taking its first deliberate steps toward shaping the world.

Stone tools, among the earliest artifacts of human ingenuity, offer a window into the minds of our ancient ancestors. The oldest known tools, discovered in East Africa, date back approximately 3.3 million years. Crafted by early hominins like Australopithecus afarensis or Kenyanthropus platyops, these tools were simple yet transformative. They consisted of sharp-edged flakes and blunt cores, created by striking one stone against another. To the modern eye, they might appear crude, but their significance lies in their intent. These tools represent a shift from reacting to the environment to actively modifying it—a hallmark of human innovation.

The emergence of more sophisticated tools coincided with the rise of Homo habilis, whose name, "handy man," reflects its association with tool use. Homo habilis refined the art of flint knapping, creating tools that were more effective for cutting meat, scraping hides, and breaking bones to access marrow. These tools, collectively known as the Oldowan industry, were not just functional; they were evidence of foresight. The ability to plan, to see a rock not as it was but as it could become, signaled a significant cognitive leap.

But stone tools were only the beginning. Around 1.76 million years ago, a new technological tradition emerged: the Acheulean industry, associated with Homo erectus. Acheulean tools, characterized by their symmetrical handaxes, were more complex and required greater skill and precision to produce. These tools reflected an increasing understanding of form and function, as well as the ability to pass down knowledge across generations. The spread of Acheulean tools across Africa, Europe, and Asia is a testament to the adaptability and mobility of Homo erectus.

Yet, no tool had a greater impact on human evolution than fire. The control of fire, achieved at least 1 million years ago and possibly earlier, was a transformative moment. Fire provided warmth, protection, and a means to cook food, unlocking new nutritional possibilities. Cooking, in particular, was a game-changer. It made food easier to chew and digest, allowing our ancestors to extract more energy from their meals. This energy surplus likely fueled the growth of the brain, an organ that consumes a disproportionate amount of calories.

Evolution and Creation Different Questions, One Path

Fire also altered social dynamics. Unlike tools, which could be used individually, fire was inherently communal. A fire required maintenance and protection, encouraging cooperation and communication among group members. The flickering light of a shared fire created a space for storytelling, bonding, and the exchange of ideas—a precursor to the complex social structures of later humans. Fire was not just a tool; it was a gathering point, a source of connection that began to define what it meant to be human.

The dual innovations of stone tools and fire did more than meet immediate needs; they reshaped the trajectory of human evolution. Tools and fire expanded the habitats our ancestors could occupy, enabling them to venture into colder climates and more challenging environments. They allowed hominins to become more efficient hunters and foragers, reducing their dependence on raw instinct and increasing their reliance on learned behaviors. This shift marked the dawn of culture—knowledge passed down not through genetic inheritance but through teaching and imitation.

These technological advances also had a profound impact on the human body. The use of tools and fire influenced the development of smaller teeth, weaker jaws, and shorter digestive tracts, as the physical demands of processing raw food diminished. At the same time, the hands and brain evolved to better handle the precise manipulation required for toolmaking and fire maintenance. This feedback loop—where technology influenced anatomy, which in turn enabled further technological advances—became a defining feature of human evolution.

Yet, it is crucial to remember that these innovations were not inevitable. They were the result of countless small experiments, of trial and error, of curiosity and necessity. Each chipped stone, each spark struck, represents a moment of creativity and perseverance. These early technologies did not emerge fully formed but evolved over hundreds of thousands of years, shaped by the environments and challenges faced by our ancestors.

The mastery of tools and fire was not just a survival strategy; it was a declaration. It signaled a departure from being passive participants in nature to becoming active agents of change. With tools and fire, humanity began to assert control over its surroundings, setting the stage for the agricultural and industrial revolutions that would follow. These early steps may seem distant, but they are the foundation upon which the modern world is built.

As we hold a smartphone or flick a light switch, it is easy to forget the profound journey that began with a sharp-edged flint or a glowing ember. But the story of stone tools and fire is embedded in every act of creation and discovery. It reminds us that innovation is not just about technology; it is about imagination, collaboration, and the unyielding drive to transform the world. In the chipped stones and the ancient hearths of our ancestors, we see not just the origins of technology but the origins of ourselves.

Evolution and Creation Different Questions, One Path

Our Lost Cousins and the Bonds We Share

The story of humanity is not a solitary journey. For much of our existence, Homo sapiens shared the Earth with other human species, close relatives who walked similar paths yet left behind divergent legacies. Among these were the Neanderthals and Denisovans, two species whose fates have become inextricably linked with our own. Though they vanished tens of thousands of years ago, their presence lingers—in our DNA, in the archaeological record, and in the tantalizing traces of shared existence that invite us to reflect on the nature of what it means to be human.

The Neanderthals, or Homo neanderthalensis, emerged approximately 400,000 years ago, their evolution paralleling the rise of Homo sapiens in Africa. Adapted to the frigid climates of Ice Age Europe and western Asia, Neanderthals were robustly built, with stocky bodies, prominent brow ridges, and larger brains than modern humans. For centuries, they were caricatured as brutish and unsophisticated, a stereotype born more of prejudice than evidence. Yet, the archaeological record reveals a different story—one of creativity, adaptability, and emotional depth.

Neanderthals were skilled hunters and toolmakers. They crafted sophisticated stone tools, built shelters, and used fire to cook their food. Fossilized remains show evidence of healed injuries, suggesting they cared for their sick and injured—a sign of social bonds that extended beyond mere survival. Perhaps most strikingly, Neanderthals engaged in symbolic behavior. They adorned themselves with jewelry made from shells and animal bones, used pigments to create body art, and possibly even buried their dead with rituals that hint at a spiritual awareness. These behaviors blur the line between "us" and "them," challenging the notion that symbolic thought is the sole domain of Homo sapiens.

The Denisovans, in contrast, remain more enigmatic. Identified primarily through DNA analysis of a few fossil fragments—teeth, a finger bone, and a jaw—found in the Denisova Cave in Siberia, Denisovans represent a mysterious branch of the human family tree. They lived across Asia and interbred with both Neanderthals and Homo sapiens, leaving genetic imprints that persist in modern populations. For example, certain genetic traits in Tibetans, which enable them to thrive at high altitudes, can be traced back to Denisovan ancestors. These fragments of evidence suggest a species that was widespread and influential, yet whose story is only beginning to emerge.

The relationship between Homo sapiens, Neanderthals, and Denisovans was complex, marked by both competition and collaboration. Genetic studies have shown that interbreeding occurred multiple times over tens of thousands of years, resulting in a shared genetic legacy. Today, most non-African humans carry between 1% and 2% Neanderthal DNA, while populations in Asia and Oceania also bear traces of Denisovan ancestry. These genetic contributions influence not only our physical traits but also aspects of our immune systems, highlighting the enduring impact of these ancient encounters.

Yet, the coexistence of these species raises profound questions. Why did Homo sapiens survive while Neanderthals and Denisovans disappeared? The answer is likely multifaceted. Environmental changes, such as the glacial cycles that reshaped the landscapes of Europe and Asia, may have placed insurmountable pressures on Neanderthal and Denisovan populations. Competition with Homo sapiens, who possessed advanced tools, complex social networks, and a capacity for rapid adaptation, likely played a role as well. However, extinction does not necessarily imply inferiority. The Neanderthals and Denisovans were remarkably successful species, thriving for hundreds of thousands of years—far longer than Homo sapiens has existed to date.

Their disappearance is not a simple narrative of replacement but one of absorption. The genetic evidence of interbreeding tells a story of connection rather than conquest, of shared lives and mingled futures. These ancient encounters shaped the course of human evolution, leaving a legacy that persists not only in our DNA but in the broader tapestry of human history. Through them, we come to see that humanity is not a single lineage but a mosaic, a blend of traits and influences that spans multiple species.

The story of Neanderthals and Denisovans also invites us to reconsider our definitions of humanity. Were they "human" in the way we understand the term? If we judge humanity by the capacity for creativity, social bonds, and adaptability, the answer is undoubtedly yes. They were not "other" but part of the same broader experiment of life, exploring different paths within the shared framework of the genus Homo. Their extinction is a reminder of the fragility of existence, a cautionary tale of the forces—environmental, biological, and cultural—that shape the fate of species.

But their legacy endures. Each fossil unearthed, each genome sequenced, deepens our understanding of these lost cousins and their contributions to the human story. They were not merely precursors to Homo sapiens but collaborators in the shaping of what we have become. Through them, we glimpse the richness of our shared past, a past that is not solely ours but one built on the lives and legacies of those who came before.

To study Neanderthals and Denisovans is to see the interconnectedness of life and the shared history that unites us across time and species. Their story is a chapter in the broader narrative of evolution, one that reminds us that humanity's rise was not inevitable but the result of countless contingencies and connections. It is a story that continues to unfold, written in the bones, genes, and artifacts they left behind. Through them, we see not just the past but the enduring threads that bind us to it—a testament to the complexity and continuity of life.

Evolution and Creation Different Questions, One Path

The Emergence of Abstract Thought and Language

For millions of years, the story of human evolution unfolded slowly, marked by incremental changes in anatomy and behavior. Then, some 70,000 years ago, something extraordinary happened—a transformation so profound that it set Homo sapiens apart not only from other species but even from its closest relatives. This shift, often referred to as the Cognitive Revolution, heralded the dawn of abstract thought and language, two defining characteristics that reshaped the trajectory of our species. With this leap, Homo sapiens transitioned from being just another clever primate to becoming a species capable of imagining, creating, and transforming its environment in ways previously unthinkable.

Abstract thought—the ability to conceptualize things beyond immediate sensory experience—marks a pivotal turning point in human evolution. Before this leap, early humans were undoubtedly intelligent, capable of crafting tools, using fire, and navigating complex social hierarchies. However, their cognitive processes were largely tied to the tangible and immediate. Abstract thought introduced a new dimension: the ability to envision what does not yet exist, to plan for the future, and to assign meaning to symbols and ideas. It was the foundation upon which art, religion, science, and culture would be built.

The archaeological record provides glimpses into this transformative period. Around 70,000 years ago, Homo sapiens began creating objects that served no immediate practical purpose—decorative beads, carved figurines, and ochre patterns etched onto rocks. These artifacts, scattered across sites from Africa to Europe, suggest the birth of symbolic thinking. They represent more than creative expression; they are evidence of a mind capable of assigning value and meaning to the intangible. A bead is not just a bead—it is a symbol, a marker of identity, status, or belief.

Language, perhaps the most significant byproduct of this cognitive revolution, amplified the power of abstract thought. Other animals communicate, and some, like primates and cetaceans, do so with remarkable sophistication. But the language of Homo sapiens was different. It was infinitely flexible, capable of conveying not just immediate needs or warnings but complex ideas, hypothetical scenarios, and shared myths. Language enabled humans to collaborate on an unprecedented scale, binding individuals into cohesive groups that could achieve together what no single individual could accomplish alone.

Evolution and Creation Different Questions, One Path

The origins of language remain one of the great mysteries of human evolution. Fossil evidence offers limited clues, as the soft tissues of the larynx and tongue do not fossilize. However, anatomical features like the position of the hyoid bone and the size of the brain's Broca's area suggest that early Homo sapiens were physiologically capable of speech. But language is more than a biological capacity—it is a cultural innovation. It required not just the ability to produce sounds but the shared agreement on what those sounds meant. This shared understanding transformed language into a tool for building and maintaining social bonds, for passing down knowledge, and for imagining collective futures.

The advent of language had profound implications for the survival and success of Homo sapiens. It enabled the transfer of knowledge across generations, allowing each new cohort to build on the achievements of its predecessors. Through language, humans could describe landscapes they had never seen, plan hunts with unparalleled precision, and transmit complex techniques for toolmaking. Language became the backbone of culture, an invisible thread connecting individuals to their communities and to the broader narrative of their species.

Perhaps most importantly, language allowed for the creation of shared myths—stories and beliefs that bound people together in ways that transcended kinship or proximity. These myths, whether religious, political, or cultural, became the foundation of large-scale societies. A tribe of 50 individuals could function through direct interaction, but to unite hundreds, thousands, or millions required shared narratives—beliefs in gods, nations, or currencies that existed only in the collective imagination. Through these shared fictions, Homo sapiens became capable of organizing at a scale that no other species could match.

The leap into abstract thought and language did not come without cost. The ability to imagine also brought the capacity to worry, to fear, and to hope. It introduced humans to existential questions: What happens after death? Why are we here? These questions, unanswerable yet deeply compelling, gave rise to religion and philosophy, as Homo sapiens sought to make sense of a world that their newly awakened minds could perceive but not fully understand. In this sense, the Cognitive Revolution was not just a biological or cultural shift but a profound existential awakening.

What caused this leap remains a topic of debate. Some scientists point to genetic changes that may have rewired the brain, while others highlight environmental pressures that favored more complex social structures and communication. The exact trigger may never be known, but its effects are undeniable. Homo sapiens did not simply evolve physically during this period; they transformed mentally and socially, becoming a species uniquely equipped to shape its own destiny.

The Cognitive Revolution also set Homo sapiens on a path of unparalleled dominance. By harnessing the power of abstract thought and language, humans could outcompete other species, including their Neanderthal and Denisovan cousins. They developed strategies, tools, and alliances that allowed them to thrive in environments as diverse as the African savanna and the frozen tundra. But this dominance came at a cost to the natural world, as humans began altering ecosystems on a scale never before seen.

The emergence of abstract thought and language is not merely a chapter in the history of Homo sapiens—it is the foundation of everything that followed. From the first carved figurine to the first spoken word, these innovations created a new kind of being: one capable of imagining the infinite and transforming the finite. The leap was not inevitable, nor was it guaranteed. But once it occurred, it set humanity on a course that would redefine not just life on Earth but the very nature of existence. Through abstract thought and language, Homo sapiens became both creator and storyteller, weaving a narrative that continues to unfold.

Kendir Ramiz

The Brain Revolution That Made Humans Unique

The human brain is, perhaps, the most extraordinary organ in the natural world. Weighing just over three pounds, it consumes 20% of the body's energy and yet holds the capacity to imagine the cosmos, compose symphonies, and contemplate its own existence. The story of Homo sapiens is, in many ways, the story of this remarkable organ—a story of its evolution, its potential, and the transformation it wrought on the species that carried it. The so-called "brain revolution" that occurred in Homo sapiens not only set them apart from other animals but also redefined the possibilities of life on Earth.

The brain of Homo sapiens is not the largest in the animal kingdom—sperm whales and elephants surpass it in size. Nor is it the most energy-efficient. What makes it unique is its complexity, particularly in the neocortex, the region responsible for higher-order functions like reasoning, language, and imagination. Over the course of human evolution, the brain underwent significant

Evolution and Creation Different Questions, One Path

expansion, growing from the modest cranial capacity of early hominins to the 1,400 cubic centimeters characteristic of modern humans. But sheer size alone does not explain the cognitive leap that distinguished Homo sapiens from their ancestors and contemporaries. The real revolution lay in the intricate wiring of neural networks, the density of connections, and the unprecedented ability to integrate and process information.

This transformation did not happen overnight. The brain of Homo habilis, an ancestor that lived over two million years ago, was already larger than that of other primates, marking the beginning of a trend. Homo erectus, with its more advanced tools and social behaviors, demonstrated further brain development. Yet, it was only with the emergence of Homo sapiens that the brain achieved a level of sophistication that could support abstract thought, self-awareness, and creativity on a scale never seen before.

One of the most significant developments in this process was the evolution of self-awareness—the ability to reflect on one's own thoughts, emotions, and existence. This awareness gave rise to a new kind of cognition, one that could transcend the immediate demands of survival. Humans began to think not only about "how" to accomplish tasks but also about "why" they mattered. This shift in perspective allowed Homo sapiens to imagine possibilities, weigh consequences, and pursue goals that extended far beyond their immediate needs.

With self-awareness came the emergence of empathy and the ability to understand the thoughts and feelings of others. This capacity for "theory of mind" transformed social interactions, enabling humans to build complex societies based on trust, cooperation, and shared goals. It was no longer enough to act instinctively or to rely solely on kinship bonds; Homo sapiens could now form alliances, negotiate conflicts, and coordinate activities with unparalleled precision. This social intelligence became a cornerstone of human success, allowing groups to outcompete other species, including their Neanderthal cousins, despite lacking physical advantages.

The brain revolution also gave rise to creativity, an ability rooted in the neural networks that connect disparate regions of the brain. This connectivity allowed Homo sapiens to combine ideas in novel ways, resulting in innovations that transformed their world. Early humans painted intricate scenes on cave walls, carved figurines from stone and ivory, and developed tools that reflected not only utility but also imagination. These creative acts were more than mere expressions of individuality; they were declarations of a mind capable of envisioning the abstract, the symbolic, and the transcendent.

Another defining feature of the brain revolution was the emergence of memory as a tool for storytelling and cultural transmission. Humans could now recall past experiences and use them to shape future decisions, creating a continuity that extended beyond individual lifespans. Oral traditions emerged, preserving knowledge, beliefs, and histories within communities. Through shared stories, Homo sapiens built a collective memory that became the foundation of culture. This cultural memory allowed each generation to build on the achievements of its predecessors, accelerating the pace of technological and social innovation.

Yet, the brain revolution was not without its costs. The cognitive abilities that set Homo sapiens apart also introduced new challenges. Self-awareness brought with it an acute understanding of mortality, a burden no other species seems to bear in the same way. The ability to imagine the future also enabled anxiety, as humans worried about threats both real and imagined. Empathy and social intelligence created opportunities for cooperation but also for manipulation, deceit, and conflict. The very traits that made Homo sapiens unique also made them vulnerable to the complexities of their own minds.

The evolutionary pressures that shaped the human brain were as much social as they were environmental. Living in groups required not only physical coordination but also the ability to navigate intricate social hierarchies and relationships. Those who could outwit rivals, charm allies, or anticipate threats were more likely to survive and reproduce. Over time, these selective pressures honed the cognitive abilities of Homo sapiens, creating a species that could adapt to nearly any environment and challenge.

What makes the brain revolution so profound is not just the capabilities it unlocked but the way it redefined what it means to be alive. For the first time, a species could contemplate its own existence, question its place in the universe, and strive to reshape its destiny. The brain gave Homo sapiens the power to transcend the limits of biology, creating tools, institutions, and ideas that extended their reach far beyond what evolution alone could achieve.

The legacy of this revolution is evident in every aspect of human life. It is there in the art that decorates the walls of caves and galleries, in the languages that carry meaning across time and space, and in the technologies that transform the world. It is there in the empathy that fosters compassion and the imagination that fuels discovery. The brain revolution was not just a step in evolution; it was a leap into a new way of being—one defined by thought, awareness, and the boundless potential of the human mind.

Through this revolution, Homo sapiens became more than a species; they became storytellers, dreamers, and creators of worlds. Their journey, shaped by the intricate dance of neurons and the spark of imagination, continues to unfold, a testament to the power of thought and the profound mystery of the human brain.

Evolution and Creation Different Questions, One Path

Creation Narratives from Sumer to Hinduism

Long before the advent of modern science, humans sought to understand their origins through stories. These narratives, woven into the fabric of ancient mythologies, served as more than just explanations—they were frameworks for meaning, offering people a sense of purpose in a world teeming with uncertainty. From the fertile plains of Mesopotamia to the spiritual traditions of the Indian subcontinent, creation myths reveal the diverse ways in which humanity has grappled with questions of existence, identity, and the cosmos.

One of the earliest and most influential mythologies emerged in ancient Sumer, a civilization that flourished in Mesopotamia around 3000 BCE. The Sumerians, among the first to develop written language, recorded their creation stories on clay tablets, preserving them for millennia. Central to their mythology was the Enuma Elish, a tale that begins in a primordial sea where the gods Apsu and Tiamat represent freshwater and saltwater, respectively. From their union emerged younger gods, whose struggles led to the creation of the world. The hero Marduk, after defeating Tiamat in a cosmic battle, used her dismembered body to form the heavens and the earth. Humanity, according to this narrative, was fashioned from the blood of a defeated god, intended to serve and sustain the divine order.

The Enuma Elish is more than a story of creation—it is a reflection of the Sumerians' worldview. It portrays a universe born from conflict and order imposed through divine will, mirroring the challenges of maintaining stability in a land prone to flooding and political upheaval. The gods were not benevolent caretakers but powerful forces to be appeased, their moods shaping the fate of humanity. This narrative reveals an early attempt to impose structure on chaos, a theme that would echo in later mythologies.

While the Sumerians imagined a world forged in battle, the Hindu tradition offers a more cyclical and philosophical perspective. Rooted in the ancient texts of the Vedas and later elaborated in the Upanishads and the Puranas, Hindu creation stories explore the interplay of creation, preservation, and destruction. Central to these narratives is the concept of Brahman, the ultimate reality that transcends time and space. Creation is not a

Evolution and Creation Different Questions, One Path

singular event but an eternal process, with the universe undergoing endless cycles of birth, death, and rebirth.

One of the most famous Hindu creation myths is found in the Rigveda, the oldest of the Vedas, composed over 3,000 years ago. The hymn known as the Nasadiya Sukta, or "Creation Hymn," begins with a profound sense of ambiguity: "There was neither non-existence nor existence then." It describes a time before creation, a void where nothingness and potential coexisted. From this formless state, a cosmic force stirred, giving rise to the universe. The hymn acknowledges the limits of human understanding, concluding with the admission that even the gods may not know the true origins of existence. This humility reflects the Hindu tradition's emphasis on mystery and the interconnectedness of all things.

Another Hindu creation narrative centers on the god Vishnu, who rests on the cosmic serpent Shesha as he dreams the universe into being. From his navel grows a lotus flower, within which sits Brahma, the creator god. Brahma then shapes the world, populating it with life. This imagery, rich with symbolism, emphasizes the cyclical and organic nature of creation. The lotus, a flower that rises from muddy waters to bloom in the sunlight, serves as a metaphor for spiritual growth and the emergence of order from chaos.

Both the Sumerian and Hindu creation stories, despite their differences, share common themes. They grapple with the tension between chaos and order, the role of divine beings in shaping the world, and humanity's place within the cosmos. These narratives reflect the societies that produced them, embedding cultural values and existential questions within their frameworks. The Sumerians, with their emphasis on hierarchy and service to the gods, mirrored the structure of their city-states, while the Hindus, with their cyclical view of time, expressed a worldview shaped by nature's rhythms.

Creation myths also serve as bridges between the sacred and the secular. They connect the material world to the divine, offering explanations for natural phenomena and legitimizing social hierarchies. In Sumer, the king was seen as a representative of the gods, his authority rooted in the divine order established through creation. In Hinduism, the idea of dharma, or cosmic law, ties individual actions to the broader workings of the universe, creating a moral framework that shapes human behavior.

What makes these myths enduring is not their literal accuracy but their ability to resonate across generations. They do not simply answer questions about how the world began; they address deeper concerns about why it exists and what it means to be human. They invite contemplation, not just of the external world but of the inner one, encouraging reflection on life's purpose and the mysteries of existence.

As we trace the evolution of creation stories, it becomes clear that they are not static. They adapt to new

Evolution and Creation Different Questions, One Path

contexts, blending with other traditions and reflecting the changing needs of the societies that uphold them. The Enuma Elish, for instance, influenced the creation narrative in the Hebrew Bible, which shares motifs such as the primordial waters and the imposition of order. Similarly, Hindu cosmology has inspired philosophical and scientific inquiries into the nature of time and the origins of the universe.

These ancient mythologies, though products of specific cultures, speak to universal human concerns. They remind us that the search for meaning is as old as humanity itself, a thread that connects us to our ancestors and to one another. Through their vivid imagery and profound questions, they invite us to see the world not as a fixed reality but as a tapestry of possibilities, shaped by the interplay of imagination and inquiry.

In exploring the creation stories of Sumer and Hinduism, we glimpse the richness of human thought and the enduring power of narrative. These myths, though ancient, remain alive in the ways they shape our understanding of the world and our place within it. They are reminders that, long before science offered its explanations, humanity sought answers through story—a testament to our capacity for wonder, reflection, and connection.

Kendir Ramiz

Humanity in the Abrahamic Faiths: A Look Beyond Adam and Eve

The story of Adam and Eve is among the most enduring narratives in human history. Rooted in the Abrahamic faiths—Judaism, Christianity, and Islam—it tells of humanity's origins in a garden of perfection, a world unmarred by suffering or sin. Yet, this story is far more than an account of the first humans. It is a profound meditation on the nature of existence, morality, and the relationship between humanity and the divine. To understand the full depth of this narrative, we must go beyond the familiar images of the Garden of Eden and the forbidden fruit, exploring how these traditions frame humanity's place in creation and the broader cosmos.

In the Book of Genesis, the story begins with the formation of the first human, Adam, from the dust of the ground. This act of creation is intimate and deliberate, as God breathes life into Adam, imbuing him with a divine spark. Adam's companion, Eve, is fashioned from his rib—a symbol often interpreted as the unity and

Evolution and Creation Different Questions, One Path

interdependence of man and woman. Together, they dwell in Eden, a paradise of abundance, tasked with stewardship over the Earth and its creatures. Their one prohibition—to refrain from eating the fruit of the Tree of Knowledge of Good and Evil—sets the stage for a dramatic turning point.

The act of defiance, often called the Fall, is a foundational moment in Abrahamic theology. It is not merely a story of disobedience but a reflection on the nature of choice and consequence. By eating the forbidden fruit, Adam and Eve gain knowledge, experiencing good and evil, shame, and mortality. This moment has been interpreted in myriad ways: as a tragic loss of innocence, as the birth of human consciousness, or as a necessary step in humanity's journey toward self-awareness and moral agency. What is clear is that this act transforms them, casting them out of Eden and into the world of struggle and survival—a world that mirrors our own.

Yet, the story of humanity in the Abrahamic traditions extends far beyond this archetypal pair. In Judaism, Adam and Eve are the ancestors of all people, but their story is not the end; it is the beginning of a covenantal relationship between God and humanity. The Hebrew Bible unfolds as a narrative of interaction between the divine and the human, exploring themes of justice, mercy, and redemption. Figures like Noah, Abraham, and Moses continue the exploration of what it means to be human, each embodying different aspects of faith, fallibility, and resilience.

Christianity, building on its Jewish roots, reinterprets the story of Adam and Eve through the lens of Jesus Christ. The Fall becomes a prelude to redemption, with Christ portrayed as the "new Adam" who restores what was lost in Eden. This theological framework emphasizes grace and salvation, casting humanity not as inherently doomed but as capable of reconciliation with the divine. The Christian narrative reframes the human condition as a journey from sin to redemption, with the promise of eternal life as a restoration of Edenic harmony.

Islam, too, offers its own perspective on Adam and Eve, emphasizing their role as the first humans and the progenitors of humanity. In the Qur'an, Adam is honored as a prophet, entrusted with knowledge and responsibility. The story of their disobedience is present, but the tone is less punitive and more didactic. Adam and Eve's repentance is accepted, underscoring God's mercy and forgiveness. In Islamic theology, their story is not a narrative of original sin but a reminder of humanity's potential for error and redemption—a cycle that defines the human experience.

What unites these traditions is their portrayal of humanity as both flawed and exalted. Humans are depicted as beings of great potential, endowed with reason, creativity, and moral awareness, yet burdened by weakness and prone to transgression. This duality lies at the heart of the Abrahamic vision of humanity. In these narratives, humans are not mere creations but

Evolution and Creation Different Questions, One Path

partners in a divine drama, their actions carrying weight and significance in a cosmos imbued with purpose.

The story of Adam and Eve also serves as a lens through which to explore humanity's relationship with the natural world. In Genesis, Adam is tasked with naming the animals and tending the garden, reflecting an early vision of stewardship and responsibility. Yet, the Fall introduces a tension between humanity and nature, as the ground becomes cursed and survival demands toil. This tension resonates today, as we grapple with questions of environmental ethics and the balance between dominion and care. The ancient story invites us to reconsider our role as caretakers of the Earth, a responsibility as pressing now as it was in the time of its telling.

Beyond the confines of Eden, the Abrahamic traditions weave a rich tapestry of human experience. They explore themes of exile and return, struggle and triumph, faith and doubt. The stories of figures like Job, who questions the justice of his suffering, or Jonah, who flees his divine mission, reveal the complexity of human emotions and the universality of existential questions. These narratives resonate not because they provide easy answers but because they mirror the ambiguities of our own lives.

In looking beyond Adam and Eve, we see that the Abrahamic traditions are not bound by a single vision of humanity. They encompass a multitude of perspectives, each contributing to a broader understanding of what it means to be human. These traditions, though rooted in specific cultures and histories, speak to universal themes of belonging, purpose, and the search for meaning. They remind us that our stories—both individual and collective—are part of a larger narrative, one that stretches across time and space.

The creation stories of Judaism, Christianity, and Islam invite us to reflect not only on where we come from but on who we are and who we might become. They challenge us to see humanity not as a fixed state but as a dynamic process, shaped by choices, relationships, and the pursuit of the divine. Through these stories, we are called to embrace our potential, confront our limitations, and participate in the unfolding story of creation itself.

Evolution and Creation Different Questions, One Path

The 6,000-Year-Old Earth: James Ussher and the Historical Roots of Radical Beliefs

Among the many narratives that attempt to explain the origins of our world, few have been as contentious or influential as the claim that the Earth is only 6,000 years old. This idea, seemingly at odds with both modern scientific understanding and ancient mythologies, did not arise in a vacuum. It is the product of a specific historical context, shaped by theological ambition, scholarly methods, and the cultural pressures of a Europe grappling with change. To understand the enduring appeal of this belief, we must delve into the work of its most famous proponent: James Ussher, the 17th-century Archbishop of Armagh.

Ussher's claim, calculated with meticulous precision, dates the creation of the world to October 23, 4004 BCE. This declaration was not made lightly; it was the culmination of years of laborious study, during which

Ussher pored over biblical genealogies, cross-referenced historical texts, and consulted ancient chronologies. In his time, the Bible was not merely a spiritual guide—it was considered an authoritative historical record. Ussher approached it with the same rigor that modern scholars might apply to archaeological data, using its genealogies to construct a timeline that connected sacred history with secular events.

The process was as complex as it was flawed. Ussher relied heavily on the genealogies found in the Old Testament, particularly those that trace the descendants of Adam through to Noah and beyond. He assigned fixed lifespans to biblical figures, stitching them together to create an unbroken chronology. This timeline was then aligned with external sources, such as the histories of the Babylonians and Persians, to provide a framework that matched biblical events with known historical occurrences. The result was a timeline that compressed the entirety of human history, from the Garden of Eden to the reign of King David, into a few millennia.

At first glance, Ussher's calculations might seem like an exercise in scholarly curiosity. Yet, they had profound implications. By declaring the Earth to be only 6,000 years old, Ussher provided a definitive chronology that reinforced a literal interpretation of scripture. This was particularly significant in a period when the authority of the Church was being challenged by scientific discoveries and the intellectual upheavals of the Renaissance. Ussher's work served as a bulwark against these challenges, asserting the primacy of biblical truth in an era of growing skepticism.

However, the 6,000-year timeline was not without its critics, even in Ussher's day. The scientific revolution was beginning to transform humanity's understanding of the natural world, driven by figures such as Galileo Galilei and Johannes Kepler. Geologists, in particular, were uncovering evidence that the Earth was far older than Ussher's calculations suggested. The study of rock strata and fossils revealed a planet shaped by processes that unfolded over immense timescales, challenging the idea of a recent creation. These discoveries did not immediately displace Ussher's timeline, but they planted seeds of doubt that would grow in the centuries to come.

Despite mounting evidence to the contrary, Ussher's chronology remained influential, particularly in religious circles. Its endurance can be attributed, in part, to its simplicity. In a world where scientific explanations were still in their infancy and often inaccessible to the average person, the idea of a 6,000-year-old Earth provided a clear and comprehensible framework. It offered certainty in a time of uncertainty, anchoring human history to a divine narrative that resonated with deeply held beliefs.

The radical nature of Ussher's claim also reflects the broader historical context in which it arose. The 17th century was a period of intense religious and political conflict, marked by the Protestant Reformation and the Catholic Counter-Reformation. In this volatile environment, Ussher's timeline was not just a scholarly endeavor; it was a statement of theological authority. By rooting his chronology in scripture, Ussher affirmed the Protestant principle of sola scriptura—the belief that the Bible alone is the ultimate source of truth. His work became a tool in the broader struggle for religious legitimacy, used to assert the supremacy of Protestant interpretations over competing views.

Over time, the 6,000-year-old Earth became a cornerstone of Young Earth Creationism, a movement that emerged in the 19th and 20th centuries as a reaction against the rise of Darwinian evolution and modern geology. Proponents of this belief clung to Ussher's chronology as a symbol of resistance to scientific secularism, framing the debate as a struggle between faith and reason. This framing, however, obscures the complexities of the historical relationship between religion and science. Figures like Ussher were not anti-science; they were deeply engaged with the intellectual tools of their time, seeking to reconcile their faith with emerging knowledge. It is ironic, then, that his work has come to represent a rejection of the very methods he employed.

The persistence of the 6,000-year timeline raises important questions about the nature of belief and its relationship to evidence. Why do ideas that conflict with overwhelming scientific consensus endure? Part of the answer lies in the human need for certainty and identity. Creation stories, whether ancient myths or modern interpretations, serve not only to explain the world but also to anchor individuals within a larger narrative. Ussher's timeline, with its definitive dates and divine origins, provides a sense of order and purpose that resonates deeply, even in the face of contradictory evidence.

Yet, Ussher's legacy is not just one of controversy. His work reminds us of the power of narrative to shape how we see the world. The idea of a 6,000-year-old Earth may no longer hold sway in scientific circles, but it continues to influence cultural and religious discourse. It challenges us to consider the ways in which stories—whether mythological, theological, or scientific—shape our understanding of reality. And it underscores the importance of questioning not just what we believe, but why we believe it.

In tracing the history of Ussher's claim, we see a microcosm of the broader human quest for knowledge—a quest that is as much about meaning as it is about facts. His story is a reminder that the search for truth is never static; it evolves alongside our understanding, shaped by the interplay of faith, reason, and imagination. Whether we view the Earth as 6,000 years old or 4.5 billion, the questions that Ussher sought to answer remain timeless: Where do we come from? And what is our place in the unfolding story of creation?

Kendir Ramiz

Faith Meets Evolution: Theistic Evolution and the Perspectives of Francis Collins

For centuries, the relationship between science and faith has been framed as a battleground—a zero-sum conflict where one must triumph over the other. Yet, this dichotomy is not only simplistic but also deeply misleading. Throughout history, many thinkers have sought to reconcile these two realms, not by diluting their essence but by finding a deeper harmony within their intersections. One of the most compelling contemporary approaches to this reconciliation is theistic evolution, a framework that integrates evolutionary science with the belief in a divine creator. Among its most prominent advocates is Francis Collins, the renowned geneticist and former director of the National Institutes of Health, whose work bridges the worlds of empirical inquiry and spiritual reflection.

At the heart of theistic evolution lies a simple yet profound proposition: that the processes described by evolutionary biology are not in conflict with the existence of a creator but are, in fact, expressions of that creator's will. It posits that natural selection, genetic mutation, and the vast expanse of time required for evolution to unfold are the mechanisms through which life's diversity has

Evolution and Creation Different Questions, One Path

been brought into being. For adherents of this view, evolution is not a challenge to belief but a testament to the intricacy and majesty of the divine design—a system so elegant that it can unfold autonomously while remaining imbued with purpose.

Francis Collins embodies this perspective in both his scientific achievements and his personal philosophy. As the leader of the Human Genome Project, Collins oversaw one of the most ambitious endeavors in modern science: the mapping of the complete human genetic code. This monumental project revealed the shared genetic heritage of all life on Earth, providing powerful evidence for evolution's explanatory power. Yet, for Collins, these discoveries did not diminish his faith; they deepened it. He famously described DNA as "the language of God," a phrase that captures his belief that the intricacies of genetics are a reflection of a divine mind.

In his book, The Language of God: A Scientist Presents Evidence for Belief, Collins outlines his journey from atheism to faith, a journey shaped by his exposure to both the natural world and the moral questions that science alone could not answer. Central to his worldview is the conviction that faith and reason are not adversaries but complementary ways of understanding reality. Science, he argues, addresses the "how" of existence—how stars form, how species evolve, how cells divide—while faith grapples with the "why"—why there is something rather than nothing, why humans seek meaning, why moral imperatives resonate across cultures.

Theistic evolution, as articulated by Collins and others, challenges the narrative that belief in evolution necessitates the rejection of God. Instead, it invites believers to see evolution as a means by which a creator has chosen to shape the world, a dynamic and ongoing process that reveals both the complexity of life and the underlying order of the cosmos. This perspective has profound implications, not only for the relationship between science and religion but also for how we understand humanity's place in the universe.

One of the key contributions of theistic evolution is its ability to reconcile seemingly contradictory narratives about human origins. Evolutionary science tells us that Homo sapiens arose through a gradual process of change, shaped by environmental pressures and genetic variation. Theistic evolution embraces this narrative while affirming that humanity's emergence is not random or purposeless. For Collins, the evolutionary process is not a series of accidents but a journey imbued with divine intention, culminating in a species capable of self-awareness, moral reflection, and spiritual longing.

This view also provides a framework for interpreting ancient religious texts in light of modern knowledge. Collins, for example, sees the biblical story of Adam and Eve not as a literal account of human origins but as a symbolic exploration of the human condition. The Fall, in this interpretation, represents humanity's capacity for moral failure and the tension between free will and divine guidance. By approaching scripture with an understanding of metaphor and allegory, theistic evolution allows for a reading that respects both the sacredness of the text and the discoveries of science.

Critics of theistic evolution often argue that it is a compromise, a way of avoiding the hard questions by blending two incompatible worldviews. But for its proponents, it is not a compromise but an expansion—a way of embracing the full scope of human understanding. It acknowledges that science and religion address different aspects of reality and that both can offer insights that enrich the other. Theistic evolution challenges the idea that belief and evidence must exist in separate domains, suggesting instead that they can inform and illuminate each other.

Collins's advocacy for theistic evolution extends beyond intellectual arguments. As a public figure and scientist, he has sought to build bridges between communities often divided by mistrust. In founding BioLogos, an organization dedicated to promoting harmony between science and faith, Collins has created a platform for dialogue, encouraging believers to engage with scientific discoveries without fear and inviting scientists to consider the spiritual dimensions of their work. BioLogos represents a vision of unity, one in which the search for truth transcends disciplinary boundaries.

The appeal of theistic evolution lies in its ability to hold complexity. It does not demand that believers abandon their faith or that scientists forsake their evidence. Instead, it invites both to see the world through a broader lens, one that acknowledges the vastness of the unknown and the interconnectedness of all things. It is a worldview that embraces mystery, finding wonder in both the precision of a genetic code and the depth of a sacred text.

In exploring the ideas of Francis Collins and theistic evolution, we are reminded that the search for meaning is not confined to any single tradition or discipline. It is a journey that spans cultures, epochs, and ways of knowing, united by a shared curiosity about the origins and purpose of existence. Theistic evolution does not claim to have all the answers, but it offers a way of asking questions that honors both the rigor of science and the transcendence of faith.

The meeting of evolution and belief is not an end point but a starting place—a space where the mysteries of life and the cosmos invite us to reflect, imagine, and wonder. It challenges us to see beyond the divisions that so often define our world, to recognize that the pursuit of truth, in all its forms, is a testament to the remarkable capacities of the human mind and spirit. Through this lens, evolution is not just a scientific theory; it is a story of creation, one that continues to unfold with every discovery and every moment of awe.

Evolution and Creation Different Questions, One Path

A Philosophical Approach: The Shared Roots of Evolution and Creation

For much of human history, the concepts of creation and evolution have been framed as opposing forces, locked in a struggle to define the origins of life and the universe. Creation, rooted in theological traditions, speaks of a divine act that brings order to chaos and life to the void. Evolution, grounded in scientific observation, describes a gradual process shaped by chance and necessity. At first glance, these narratives appear irreconcilable—one guided by purpose, the other by randomness. Yet, when approached through a philosophical lens, they reveal surprising commonalities, shared roots that point to a deeper unity in how humans seek to understand existence.

The tension between creation and evolution is, in many ways, a reflection of the human desire for certainty. Creation myths, found across cultures and epochs, offer definitive answers to profound questions: Where did we come from? Why are we here? These narratives provide structure, anchoring humanity within a broader cosmic order. Evolution, on the other hand, embraces ambiguity, offering a model that is constantly revised as new evidence emerges. It is less a story with a fixed ending and more a process—a dynamic unfolding that mirrors the unpredictability of life itself. This contrast can make the two seem fundamentally at odds. But what if this tension is itself a product of limited perspective?

To explore the shared roots of creation and evolution, we must first step back and examine their underlying assumptions. At their core, both are attempts to grapple with the same mystery: the emergence of complexity from simplicity. Whether one speaks of a divine creator shaping the cosmos or natural selection driving the diversification of life, the fundamental question remains: How does order arise from chaos? This shared focus on emergence—a process by which new structures, patterns, and forms come into being—points to a philosophical convergence that transcends the particulars of any single narrative.

Creation stories often begin with a primordial state of chaos, a formless void from which the universe is born. In the Enuma Elish, the Babylonian creation myth, the gods bring order to a chaotic sea, shaping the heavens and the earth. In the Rigveda, the Hindu tradition describes an undifferentiated state where existence and nonexistence are intertwined, from which creation springs. These narratives emphasize the transformative power of agency—whether divine or metaphysical—to impose structure on an otherwise unordered reality.

Evolution, too, is a story of emergence, though its mechanisms are rooted in natural processes rather than divine intervention. It begins not with chaos but with

Evolution and Creation Different Questions, One Path

simplicity: single-celled organisms that, over billions of years, give rise to the staggering diversity of life. Evolution demonstrates that complexity can arise not through intent but through the accumulation of small, incremental changes—an elegant dance of variation, selection, and inheritance. This process, while lacking the conscious agency of creation myths, is no less extraordinary. It reveals a universe capable of generating beauty and complexity through its own intrinsic laws.

What connects these narratives is not just their focus on emergence but their shared recognition of interconnectedness. Creation stories often depict humanity as intimately linked to the cosmos, shaped from the same elements that constitute the stars and the earth. In the Book of Genesis, Adam is formed from the dust of the ground, emphasizing the continuity between humans and the natural world. Evolutionary science echoes this sentiment, showing that all life on Earth shares a common ancestry. The genetic code that defines Homo sapiens is written in the same language as that of bacteria, birds, and blue whales. In both narratives, humanity is not separate from nature but an integral part of it.

This interconnectedness invites a rethinking of the supposed conflict between creation and evolution. Rather than opposing forces, they can be seen as complementary perspectives, each illuminating different aspects of the same phenomenon. Creation speaks to purpose, meaning, and the human need for narrative coherence. Evolution, with its focus on process and evidence, provides a framework for understanding the mechanisms through which life unfolds. Together, they form a more complete picture—one that acknowledges both the wonder of existence and the intricacies of its unfolding.

Philosophers throughout history have grappled with the relationship between purpose and process, seeking ways to bridge the gap between these two ways of seeing the world. The ancient Greek philosopher Heraclitus spoke of a cosmos governed by logos, a rational principle that underlies the apparent chaos of existence. In the 20th century, Teilhard de Chardin, a Jesuit priest and paleontologist, proposed a vision of evolution imbued with spiritual significance. For Teilhard, evolution was not merely a biological process but a cosmic one, moving toward greater consciousness and unity—a view that sought to harmonize science and faith.

These philosophical approaches remind us that creation and evolution are not mutually exclusive. Instead, they offer different tools for exploring the same mystery. Creation asks "why," seeking meaning and purpose. Evolution asks "how," uncovering the mechanisms and patterns that shape life. Both are products of the human imagination and intellect, driven by the same desire to make sense of the world.

Understanding the shared roots of creation and evolution also has profound implications for how we

Evolution and Creation Different Questions, One Path

view ourselves. Creation narratives often emphasize humanity's uniqueness, portraying us as the pinnacle of divine intention. Evolution, by contrast, situates us within a broader web of life, one species among many, shaped by the same forces that govern all living things. Yet, these perspectives are not contradictory. They can coexist, offering a vision of humanity that is both humble and extraordinary. We are creatures of the Earth, bound by its laws, and yet capable of contemplating the universe and our place within it.

To reconcile creation and evolution is not to diminish either but to expand our understanding of both. It is to recognize that the questions they seek to answer—questions of origin, purpose, and interconnectedness—are larger than any single narrative. It is to embrace complexity, to see the beauty in ambiguity, and to acknowledge that the search for truth is as much about the journey as it is about the destination.

Through a philosophical lens, creation and evolution are not adversaries but partners in a shared quest—a quest to understand not only where we come from but what it means to be alive. They remind us that the universe is vast, mysterious, and deeply interconnected, and that our attempts to comprehend it, whether through story or science, are a testament to the boundless creativity of the human mind. In this interplay of faith and reason, we find not conflict but harmony, a reflection of the same unity that underlies the cosmos itself.

Kendir Ramiz

TheSeverance from Nature: Humanity's Ambition to Control the Environment

For the vast majority of human history, our species lived in harmony with nature, not by choice but by necessity. Early Homo sapiens were deeply attuned to the rhythms of the natural world, relying on their environment for sustenance, shelter, and survival. They hunted, gathered, and adapted to the shifting seasons, their lives shaped by forces far beyond their control. But over time, something changed. Humanity began to pull away from the natural order, no longer content to exist within its bounds but determined to reshape it. This drive to control the environment, while a hallmark of our species' ingenuity, also marked the beginning of a profound disconnection from the natural world.

Evolution and Creation Different Questions, One Path

The roots of this severance can be traced back to a single, defining trait of Homo sapiens: the capacity to imagine alternative futures. Unlike other animals, which adapt to their environments through instinct and incremental evolution, humans could envision possibilities beyond the immediate present. This ability allowed us to innovate, to experiment, and to transform our surroundings in ways no other species could. It was a gift that enabled survival in harsh climates and the development of tools, fire, and shelter. But it also planted the seeds of an ambition that would come to define our species—the desire to dominate the natural world rather than coexist with it.

The Neolithic Revolution, which began roughly 10,000 years ago, marked a turning point in humanity's relationship with nature. With the advent of agriculture, humans shifted from a nomadic existence to one rooted in the cultivation of land. This transition brought unprecedented stability and surplus, enabling the rise of villages, cities, and civilizations. But it also fundamentally altered the dynamics of human-environment interaction. The land, once a source of sustenance to be respected, became a resource to be exploited. Forests were cleared, rivers diverted, and ecosystems reshaped to serve human needs. This transformation, while revolutionary, came at a cost—both to the natural world and to the balance that had sustained humanity for millennia.

The ambition to control nature extended beyond agriculture. As civilizations grew, so did their technological capabilities. Irrigation systems, roads, and fortifications allowed humans to thrive in environments that had once seemed inhospitable. In ancient Mesopotamia, the construction of canals turned arid landscapes into fertile fields, while the pyramids of Egypt and the aqueducts of Rome stood as monuments to humanity's ability to bend nature to its will. Yet, these achievements often masked the fragility of the systems they created. The salinization of soils, deforestation, and the depletion of resources foreshadowed the long-term consequences of humanity's growing dominance over the natural world.

This trajectory reached new heights with the Industrial Revolution. Fueled by coal and steam, humanity entered an era of unprecedented power over nature. Factories belched smoke into the sky, railways crisscrossed continents, and cities expanded at a breakneck pace. The natural world, once a source of wonder and reverence, became a backdrop to human progress—valued not for its intrinsic worth but for its utility. Forests were felled, rivers polluted, and species driven to extinction, all in the name of economic growth and technological advancement. The Earth, once a partner in humanity's journey, was now seen as a commodity to be consumed.

At the heart of this disconnection lies a paradox. The very qualities that enabled humanity's rise—creativity, adaptability, and ambition—also led to an estrangement from the natural systems that sustain life. By seeking to control nature, humans distanced themselves from it,

Evolution and Creation Different Questions, One Path

forgetting that their survival depended on the delicate balance of ecosystems. This illusion of dominance has been reinforced by cultural narratives that place humanity above and apart from the natural world. From the biblical command to "subdue the Earth" to the Enlightenment ideal of conquering nature through reason, these stories have shaped a worldview in which progress is measured by the degree to which humans can overcome natural limitations.

Yet, the consequences of this mindset are becoming increasingly apparent. Climate change, biodiversity loss, and environmental degradation are not just challenges for the future—they are crises unfolding in the present. The same ingenuity that enabled humans to harness the power of nature now threatens to destabilize it. Rising sea levels, extreme weather events, and the collapse of ecosystems serve as stark reminders that humanity's control over nature is not absolute. Instead, it is a precarious illusion, built on systems that are both interconnected and vulnerable.

The severance from nature is not merely an environmental issue; it is a psychological and spiritual one. As humans have distanced themselves from the natural world, they have also lost something intangible—a sense of belonging to a larger whole. Indigenous cultures, which often view humanity as an integral part of nature rather than its master, offer a stark contrast to the dominant Western paradigm. For many indigenous peoples, the land is sacred, its rhythms and cycles woven into their identities and ways of life. This perspective challenges the notion that humanity's destiny lies in controlling nature, suggesting instead that true harmony comes from living in balance with it.

Reconnecting with nature requires more than technological solutions; it demands a fundamental shift in perspective. It calls for a recognition that humanity is not separate from the natural world but deeply embedded within it. The air we breathe, the water we drink, and the food we eat are all products of ecosystems that we influence but do not control. Acknowledging this interconnectedness is the first step toward restoring harmony—a harmony that, once lost, can only be reclaimed through humility and stewardship.

The story of humanity's severance from nature is not just a cautionary tale; it is an invitation to reimagine our relationship with the world around us. It challenges us to question the narratives of dominance and progress that have shaped our history, to seek a new balance that honors both human ingenuity and the integrity of the natural world. In this reimagining, there is hope—not for a return to the past, but for a future in which humanity and nature coexist, not as adversaries, but as partners in the shared journey of life.

Evolution and Creation Different Questions, One Path

The Flaws of Homo Sapiens: Our Physical and Genetic Weaknesses

Homo sapiens stands as the most dominant species in the history of the planet, wielding unparalleled power over the natural world. Yet, beneath this veneer of mastery lies a creature marked by profound vulnerabilities. Despite our cognitive brilliance, our physical and genetic frailties reveal a deeper truth: we are not the perfect beings that our achievements might suggest. The story of Homo sapiens is not one of flawless evolution but of adaptation, compromise, and survival in spite of our weaknesses.

Physically, Homo sapiens is a relatively unimpressive organism. Compared to many of our evolutionary cousins and other animals, we are neither particularly strong nor fast. Our skeletal structure, while enabling bipedal locomotion, comes with significant trade-offs.

Walking upright frees our hands for tool use and complex tasks, but it also places tremendous stress on our spines. Back pain, herniated discs, and poor posture are all consequences of a design optimized for movement rather than endurance. The curvature of the spine, a relatively recent adaptation in evolutionary terms, often struggles to bear the weight of a modern, sedentary lifestyle, exacerbating these issues.

Our feet, too, are imperfect. The arch that allows for efficient walking and running is prone to collapse, resulting in conditions like flat feet and plantar fasciitis. Knees, tasked with bearing the full weight of an upright body, are notoriously susceptible to injury, particularly in athletes or individuals who engage in repetitive motions. These vulnerabilities highlight the fact that our anatomy, while extraordinary in its utility, is a product of compromise rather than perfection. Evolution does not aim for flawless design; it selects for traits that are "good enough" to ensure survival and reproduction.

Internally, the human body is equally fraught with inefficiencies. The appendix, a vestigial organ left over from our herbivorous ancestors, serves little purpose today yet can become life-threatening if infected. Wisdom teeth, once essential for chewing coarse, unprocessed foods, now crowd modern jaws, often requiring surgical removal. These remnants of our evolutionary past are not merely curiosities; they are reminders that evolution is a patchwork process, building upon existing structures rather than starting from scratch.

Evolution and Creation Different Questions, One Path

Perhaps even more striking than our physical imperfections are the genetic vulnerabilities that permeate our species. The very mechanisms that enable genetic diversity and adaptation also give rise to inherited diseases and disorders. Sickle cell anemia, cystic fibrosis, and Huntington's disease are just a few examples of conditions rooted in our genetic code. While some of these mutations have historical advantages—sickle cell traits, for instance, confer resistance to malaria—their persistence underscores the precarious balance between survival and suffering.

The human genome, far from being a pristine blueprint, is riddled with errors and redundancies. Mutations occur at a rate of about 100 per individual per generation, most of which are neutral or deleterious. This genetic "noise" reflects the messy reality of evolution, where natural selection works with existing material rather than engineering perfection. Moreover, the increasing reliance on modern medicine has altered the selective pressures that once shaped our species, allowing genes that might have been culled in harsher environments to persist and propagate.

One of the most striking examples of Homo sapiens' genetic fragility is our susceptibility to disease. Unlike species that have coevolved with their environments to achieve a degree of resilience, humans have created conditions that exacerbate their vulnerabilities. Urbanization, global travel, and agriculture have facilitated the spread of infectious diseases, from smallpox to COVID-19. Our immune systems, while remarkably adaptable, are not invincible. Autoimmune disorders, allergies, and cancers—where the body turns against itself—highlight the delicate balance our biology struggles to maintain.

But perhaps the most significant flaw of Homo sapiens lies not in our physical or genetic weaknesses but in the cognitive biases that have shaped our behavior. Our brains, extraordinary as they are, are also deeply flawed. We are prone to overconfidence, short-term thinking, and tribalism—traits that once served us well in small, close-knit communities but now hinder our ability to address global challenges. The same cognitive leaps that allowed us to dominate the planet also burden us with anxiety, existential dread, and a propensity for self-destruction.

These imperfections, however, are not reasons to despair. On the contrary, they underscore the resilience and adaptability that define our species. Homo sapiens' ability to thrive despite these flaws is a testament to the ingenuity and collaboration that have driven our evolution. Medicine, technology, and social structures have allowed us to mitigate many of our vulnerabilities, extending lifespans and improving quality of life in ways unimaginable to our ancestors.

Yet, our imperfections also serve as a humbling reminder of our place within the broader web of life. We are not the pinnacle of evolution but one branch among many, shaped by the same forces that govern all living things. Our flaws are not failures; they are the evidence of a process that values survival over perfection, adaptability over design. They connect us to our evolutionary past and challenge us to confront the limitations of our present.

In recognizing our physical and genetic weaknesses, we are invited to approach our existence with a sense of wonder rather than hubris. These vulnerabilities are not just markers of what we lack but also of what we have overcome. They remind us that our dominance is not guaranteed, that our future depends on our ability to adapt—not just biologically but socially, ethically, and environmentally. The story of Homo sapiens is one of imperfection and resilience, a testament to the power of life to persist and evolve in the face of adversity.

Kendir Ramiz

Ecological Balance and Humanity: How Natural Order Was Disrupted

For most of its existence, Homo sapiens lived as a participant in the intricate web of life, adapting to the natural rhythms of ecosystems and relying on their balance for survival. Early humans hunted and gathered with reverence, understanding that their lives were intertwined with the seasons, the soil, and the creatures that shared their world. They were not masters of the Earth but part of its cycle, taking what was needed and leaving enough for the system to regenerate. Yet, as Homo sapiens evolved, so too did their relationship with nature. What began as coexistence slowly shifted to domination, disrupting ecological balance in ways that would ripple across millennia.

The concept of ecological balance is as ancient as life itself. Ecosystems, from rainforests to coral reefs, function as dynamic systems in which each species plays a role, maintaining the stability of the whole. Predators regulate prey populations; plants convert sunlight into energy; fungi recycle nutrients. It is a self-sustaining cycle, delicately tuned to the unique

Evolution and Creation Different Questions, One Path

conditions of each environment. When one element of the system falters, the consequences cascade, affecting every other component. For millions of years, this balance endured, allowing life to flourish in myriad forms.

The disruption of this balance began with a single, seemingly innocuous act: the mastery of fire. While fire provided warmth, protection, and the ability to cook food, it also gave humans the power to reshape their environments. Forests were burned to create open land for hunting, and later for agriculture. This marked the first significant human impact on ecosystems—a foreshadowing of far greater changes to come.

The Neolithic Revolution, which ushered in agriculture, was a turning point in humanity's relationship with nature. No longer dependent on the whims of the wild, humans began cultivating specific crops and domesticating animals, creating controlled ecosystems that prioritized human needs over natural diversity. This shift brought unprecedented stability and growth but came at a cost. Forests were cleared to make way for fields; wetlands were drained; rivers were diverted. Each transformation disrupted the habitats of countless species, reducing biodiversity and weakening the resilience of ecosystems.

As human populations grew, so too did their demands on the Earth. The rise of civilizations brought monumental achievements in art, science, and culture, but these came hand-in-hand with environmental degradation. Ancient Mesopotamia, for example, flourished on the fertile plains between the Tigris and Euphrates rivers, but intensive irrigation practices led to salinization of the soil, rendering large areas barren. Similarly, deforestation in ancient Greece fueled shipbuilding and urban expansion but left the landscape vulnerable to erosion and desertification. These early examples highlight a recurring pattern: humanity's tendency to prioritize short-term gains over long-term sustainability.

The Industrial Revolution magnified this trend on an unprecedented scale. Fueled by fossil energy, humans gained the ability to extract resources and alter environments at a pace unmatched in history. Forests were felled to feed the furnaces of industry; rivers were dammed to power machines; landscapes were carved apart to extract coal and minerals. The natural world, once seen as a partner, became a commodity—a source of raw materials to fuel human progress. This exploitation disrupted ecosystems on a global scale, driving many species to extinction and transforming entire regions beyond recognition.

Perhaps the most profound consequence of humanity's disruption of ecological balance is the loss of biodiversity. Every ecosystem relies on a diversity of species to function effectively. Predators maintain the health of prey populations; pollinators enable plants to reproduce; scavengers recycle nutrients. When species disappear, the system becomes less stable, less capable of adapting to changes. The extinction of a

Evolution and Creation Different Questions, One Path

single pollinator, for example, can disrupt an entire food chain, leading to declines in plant and animal populations alike.

Climate change, driven by human activity, has further destabilized ecosystems. Rising temperatures, shifting weather patterns, and melting ice caps are altering habitats at a pace too rapid for many species to adapt. Coral reefs, which support a quarter of all marine life, are dying due to ocean warming and acidification. Forests are succumbing to drought, pests, and fires. Entire ecosystems are reaching tipping points, where the balance they once maintained is irreversibly lost.

Yet, this story is not solely one of destruction. It is also a story of interdependence and the possibility of renewal. Despite the damage inflicted by Homo sapiens, the natural world retains a remarkable capacity for recovery. Forests can regrow; rivers can cleanse themselves; species can rebound—if given the chance. This resilience is a testament to the power of ecological balance, a reminder that even in the face of disruption, the natural world remains a force of regeneration.

To restore balance, however, requires a shift in perspective. Humanity must move away from viewing itself as separate from nature and instead embrace its role as a steward within it. Indigenous cultures offer valuable lessons in this regard. Many indigenous traditions emphasize the interconnectedness of all life, advocating for practices that maintain harmony rather than exploitation. From sustainable hunting methods to rotational farming, these practices reflect a deep understanding of ecological balance—a wisdom often overlooked in the modern, industrialized world.

The disruption of natural order is not merely an environmental crisis; it is a crisis of identity. In severing ties with the natural world, humanity has lost a sense of its place within the broader web of life. The air we breathe, the water we drink, the food we eat—all are gifts of ecosystems that we have pushed to their limits. To reclaim balance is not simply to protect nature; it is to protect ourselves, to recognize that our fate is inextricably linked to the health of the planet.

The story of ecological balance and its disruption is a reminder of the duality of human nature: our capacity for destruction and for renewal, for shortsightedness and for foresight. It challenges us to confront the consequences of our actions and to imagine a future where humanity and nature coexist in harmony. This future is not inevitable, but it is possible—a possibility rooted in the resilience of the natural world and the ingenuity of the human spirit. In the delicate interplay of ecosystems, there is hope, not only for the Earth but for our place within it.

Evolution and Creation Different Questions, One Path

An Evolutionary Dead End?: Why the Human Species Is at Odds with Nature

Humanity's relationship with nature is paradoxical. On one hand, Homo sapiens evolved as part of the natural world, subject to its rules and reliant on its resources for survival. On the other hand, our species has become nature's greatest disruptor, reshaping ecosystems, driving countless species to extinction, and altering the planet's climate. How did this happen? What is it about Homo sapiens that makes us so profoundly out of sync with the natural systems that gave rise to us? To understand this tension, we must explore whether our trajectory represents an evolutionary success or a dangerous deviation—a path that might ultimately lead to a dead end.

Evolution is often misunderstood as a linear march toward perfection, with each species representing an improvement over the last. But this is a misleading metaphor. Evolution is not about progress; it is about adaptation. Species evolve not to become "better" in an absolute sense but to survive in specific environments. By this measure, Homo sapiens is undoubtedly successful. We have spread to every corner of the globe, adapted to climates ranging from arctic tundras to tropical rainforests, and developed technologies that allow us to thrive even in the harshest conditions. Yet, our success comes with a caveat: it has come at the expense of the natural world, and perhaps even our own long-term survival.

The roots of humanity's disconnection from nature lie in our evolutionary journey. Unlike most species, which adapt through gradual changes in physiology, Homo sapiens developed an extraordinary cognitive capacity that allowed us to reshape our environments rather than merely adapt to them. This cognitive leap—the so-called Cognitive Revolution—enabled us to create tools, build shelters, and manipulate landscapes to suit our needs. It also gave rise to abstract thought, language, and imagination, allowing us to envision futures and develop strategies to achieve them. While these abilities were instrumental in our rise to dominance, they also planted the seeds of our estrangement from nature.

One of the defining characteristics of Homo sapiens is our ability to think beyond the immediate present. This capacity for long-term planning and goal-setting is unique among animals and has driven many of our most remarkable achievements. However, it has also led to a profound shift in how we relate to the natural world. For most of human history, early Homo sapiens lived in relative harmony with nature, relying on its resources but constrained by its limits. But as our cognitive and technological capabilities grew, so did our ambition to transcend those limits. No longer content to coexist with nature, we sought to control it.

This drive for control has led to a fundamental misalignment between humanity and the ecosystems we inhabit. Unlike other species, which exist as part of balanced food webs and ecological cycles, humans have systematically disrupted those systems to serve our own purposes. Agriculture, industrialization, and urbanization have allowed us to extract more from the Earth than any other species, but they have also decoupled us from the natural processes that sustain life. We no longer live within ecosystems; we dominate them, often with little regard for the consequences.

But why has this disconnection occurred? Some researchers suggest that the very traits that make Homo sapiens unique—our intelligence, creativity, and adaptability—are also the source of our ecological discord. These traits have enabled us to override the feedback loops that keep other species in check. Predators, for example, are limited by the availability of prey, while herbivores are constrained by the abundance of vegetation. Humans, however, have developed the ability to circumvent these limits. We domesticated plants and animals, harnessed fossil fuels, and invented technologies that allow us to extract resources at unprecedented scales. This has created a situation in which our impact on the planet is no longer regulated by natural constraints.

Another factor is the evolution of human social structures. Unlike other animals, whose behaviors are shaped primarily by genetic instincts, humans are heavily influenced by culture. This has allowed us to create complex societies with shared values, norms, and technologies. However, it has also led to the emergence of systems that prioritize short-term gains over long-term sustainability. Economic growth, for example, has become a central goal of modern societies, often at the expense of ecological health. This cultural evolution, while beneficial in some respects, has exacerbated our disconnection from nature by reinforcing the illusion that we are separate from it.

The question, then, is whether humanity's current trajectory represents an evolutionary dead end. From an ecological perspective, the answer is troubling. Many of the behaviors that have driven our success—deforestation, overfishing, and reliance on nonrenewable resources—are undermining the very systems that sustain us. Climate change, biodiversity loss, and pollution are not just environmental crises; they are existential threats, symptoms of a species that has outgrown its ecological niche. If unchecked, these trends could lead to a collapse of the systems that support human life, bringing our era of dominance to an abrupt and catastrophic end.

Yet, it is also important to recognize that evolution is not destiny. Unlike other species, which are bound by their instincts and genetic programming, Homo sapiens has the ability to reflect on its actions and change course.

Evolution and Creation Different Questions, One Path

Our cognitive flexibility—the same trait that has driven our ecological impact—also gives us the potential to adapt in ways that are unprecedented in the history of life. We can choose to align our behaviors with the principles of sustainability, to reintegrate ourselves into the natural systems we have disrupted. This will require not only technological innovation but also a fundamental shift in how we view our place in the world.

Reframing our relationship with nature means recognizing that we are not separate from it but deeply interconnected with it. The air we breathe, the water we drink, and the food we eat all depend on ecosystems that we influence but do not control. Acknowledging this interconnectedness is the first step toward restoring balance. It is not a return to a romanticized past but a reimagining of the future—one in which humanity thrives not in opposition to nature but as a part of it.

The story of Homo sapiens is still being written. Whether it ends in ecological collapse or a new era of harmony depends on the choices we make in the coming decades. What is clear is that our current path is unsustainable, a reflection of a species that has yet to fully understand the consequences of its actions. But just as evolution has shaped our past, so too can it shape our future. By embracing our capacity for reflection and change, we can redefine what it means to be human—not as conquerors of the natural world, but as its stewards, its partners, and its caretakers.

Kendir Ramiz

Reconciliation with Nature: Crafting a Future Aligned with Our Evolutionary Legacy

Homo sapiens is a species caught in a paradox. Born from the cradle of nature, shaped by millions of years of evolution, we now find ourselves estranged from the very systems that made our existence possible. Our rapid rise to dominance over the planet has come at a cost—both to the natural world and to our own sense of belonging within it. Yet, this disconnection is not inevitable. The same capacities that have allowed us to disrupt the Earth's ecosystems—our intelligence, creativity, and adaptability—also give us the power to mend our relationship with nature. The question is not whether such reconciliation is possible but whether we will choose to pursue it.

To imagine a future aligned with our evolutionary heritage, we must first understand the nature of that heritage. For most of our history, Homo sapiens lived in intimate connection with the natural world. Our ancestors hunted and gathered with a deep awareness of their environment, guided by the seasons, the

Evolution and Creation Different Questions, One Path

movements of animals, and the cycles of plants. Survival required not only physical endurance but also a profound understanding of ecosystems—knowledge passed down through generations and embedded in cultural practices. This way of life fostered a respect for the balance of nature, a recognition that humanity was part of a larger web of life.

The agricultural revolution, industrialization, and urbanization disrupted this balance, distancing humans from their evolutionary roots. Yet, these transformations did not erase our connection to nature; they merely obscured it. Beneath the layers of concrete and steel, our biology remains deeply tied to the rhythms of the natural world. Our bodies respond to sunlight and darkness, our minds find solace in green spaces, and our well-being is intertwined with the health of the ecosystems that sustain us. This evolutionary inheritance is both a reminder of what we have lost and a guide to what we might reclaim.

Creating a future that honors this inheritance begins with a shift in perspective. For too long, humanity has viewed itself as separate from nature, an exception to its rules. This mindset has driven much of the environmental degradation we see today, from deforestation to climate change. But science and experience tell a different story: we are not apart from nature but deeply embedded within it. The air we breathe, the water we drink, and the food we eat are all products of ecosystems that we influence but do not control.

Acknowledging this interconnectedness is the first step toward reconciliation.

One path to reconnection lies in rethinking our relationship with the natural world, moving from exploitation to stewardship. Indigenous cultures offer profound insights in this regard. Many indigenous traditions emphasize the interconnectedness of all life, advocating for practices that sustain rather than deplete. These practices—such as rotational farming, sustainable hunting, and reverence for natural cycles—reflect an understanding of humanity's place within the ecological web. They remind us that progress does not have to come at the expense of balance, and that innovation can be guided by respect for natural systems.

At the same time, modern science and technology hold immense potential to support this vision. Advances in renewable energy, precision agriculture, and ecological restoration provide tools to reduce humanity's environmental footprint and repair damaged ecosystems. Rewilding initiatives, which reintroduce native species and restore habitats, have demonstrated nature's capacity for resilience when given the chance. Urban design can integrate green spaces, creating cities that coexist with nature rather than displace it. These solutions, though diverse in their approaches, share a common goal: to harmonize human activity with the needs of the planet.

Yet, reconciliation is not just a matter of external changes; it also requires an internal shift. Our estrangement from nature is as much a psychological and cultural phenomenon as it is a physical one. Modern life, with its screens and schedules, often leaves little room for the natural world. But studies consistently show that time spent in nature improves mental health, reduces stress, and fosters a sense of connection. Reintroducing nature into our daily lives—whether through urban parks, community gardens, or simply walking in the woods—is not a luxury but a necessity. It reminds us of our place within the larger story of life and rekindles the wonder that so often drives us toward preservation.

Education plays a critical role in this process. Teaching future generations about the interdependence of life, the principles of ecology, and the consequences of human actions empowers them to make informed choices. Environmental education is not just about facts; it is about cultivating a sense of responsibility and possibility. By grounding children and young adults in the realities of our shared planet, we can inspire a new generation of stewards who view the natural world not as a resource to be exploited but as a community to be nurtured.

Reconciliation with nature also requires confronting the inequalities that drive environmental harm. Climate change and biodiversity loss disproportionately affect the most vulnerable populations, exacerbating social and economic divides. Addressing these issues means recognizing that environmental justice and social justice are inseparable. A future aligned with our evolutionary legacy must be one that prioritizes equity, ensuring that the benefits of ecological harmony are shared by all.

The journey toward harmony with nature is not about returning to a romanticized past; it is about envisioning a sustainable future. It is about acknowledging that our evolutionary success is intertwined with the well-being of the planet and that our destiny is inseparable from that of the ecosystems we inhabit. This vision is not an abstraction; it is a necessity, a recognition that the challenges we face—climate change, resource depletion, and biodiversity loss—are not external threats but symptoms of a deeper disconnection.

As we look to the future, the question is not whether humanity can reconcile with nature but whether it will choose to. The path forward is one of humility and hope, of recognizing both our limits and our potential. It is a path that asks us to draw on the wisdom of our evolutionary past while embracing the possibilities of our creative minds. In doing so, we have the chance to craft a future that honors not only the Earth but the intricate web of life that sustains us. This is the legacy we inherited, and it is the legacy we have the power to shape.

Evolution and Creation Different Questions, One Path

Laboratory Experiments: Insights from E. coli and Fruit Fly Studies

In the realm of science, there are few endeavors more illuminating than the study of evolution as it unfolds in real time. While the vast majority of evolutionary changes occur over millennia, far beyond the scope of human observation, laboratory experiments have provided a rare glimpse into the mechanics of natural selection and adaptation. Among the most compelling examples are long-term studies of E. coli bacteria and fruit flies (Drosophila melanogaster), organisms that have become cornerstones of evolutionary research. These experiments, though conducted in controlled environments, have revealed profound truths about the processes that drive change and diversification in life.

The E. coli long-term evolution experiment (LTEE), initiated by Richard Lenski in 1988, stands as one of the most ambitious and illuminating studies in evolutionary biology. The premise was simple yet groundbreaking: to observe how bacterial populations evolve over thousands of generations under consistent conditions. Lenski and his team began with 12 identical populations of E. coli, all derived from the same ancestral strain. These bacteria were grown in a glucose-limited medium, ensuring a constant but finite food source, which acted as a selective pressure.

Over the course of the experiment—now spanning more than 70,000 generations—E. coli populations have undergone remarkable changes. One of the most striking discoveries was the evolution of a novel trait in one of the populations: the ability to metabolize citrate, a compound present in the growth medium but typically unusable by E. coli under aerobic conditions. This adaptation, which occurred after tens of thousands of generations, demonstrated the power of mutation and natural selection to produce entirely new capabilities. Genetic analysis revealed that this innovation arose through a series of mutations, each building upon the last, highlighting the cumulative nature of evolutionary change.

The LTEE has also shed light on broader evolutionary dynamics, such as the balance between random mutation and deterministic selection. While mutations occur randomly, the consistent environment of the experiment has led to parallel adaptations across multiple populations. For example, all populations have evolved faster growth rates and larger cell sizes, traits that provide competitive advantages in their glucose-limited world. These parallels underscore the predictability of evolution under certain conditions, even as the specific genetic pathways vary.

Similarly, experiments with fruit flies have provided invaluable insights into the mechanisms of evolution. Drosophila melanogaster is an ideal model organism due to its short generation time, genetic simplicity, and ease of manipulation. Researchers have used fruit flies to study everything from genetic mutations to sexual selection, revealing the intricate interplay of forces that shape evolutionary outcomes.

One seminal series of experiments involved subjecting fruit fly populations to environmental stressors, such as changes in temperature or the introduction of toxins. Over successive generations, the flies evolved traits that enhanced their survival in these challenging conditions. For instance, populations exposed to extreme heat developed greater thermal tolerance, a change driven by mutations in heat-shock protein genes. These findings illustrate how selective pressures act on genetic variation to produce adaptations, often with remarkable speed.

Another fascinating area of research has focused on genetic drift, the random changes in gene frequency that occur in small populations. By isolating small groups of fruit flies and tracking their genetic changes over generations, scientists have observed how chance events, rather than selection, can shape evolutionary trajectories. These studies have provided a nuanced understanding of how evolution operates in populations of varying sizes, highlighting the interplay between randomness and adaptation.

Laboratory experiments with E. coli and fruit flies have also illuminated the concept of fitness landscapes, a metaphorical representation of how different genetic combinations influence an organism's reproductive success. Evolutionary biologists often describe these landscapes as rugged, with peaks representing optimal adaptations and valleys corresponding to less advantageous traits. The LTEE has demonstrated how populations navigate these landscapes, sometimes climbing directly toward a peak and other times becoming trapped on suboptimal plateaus. This process reflects the constraints and opportunities inherent in evolutionary change, as well as the role of historical contingency in shaping outcomes.

The implications of these studies extend far beyond the laboratory. They provide empirical evidence for the core principles of evolution, validating theories that were once purely speculative. Moreover, they underscore the adaptability of life, illustrating how even simple organisms like bacteria and fruit flies can innovate and thrive in the face of changing conditions. These findings resonate with broader questions about the resilience of ecosystems and the capacity for life to adapt to the challenges posed by human activity, such as climate change and habitat destruction.

Perhaps most importantly, these experiments remind us that evolution is not a distant, abstract process confined to the fossil record or the distant past. It is an ongoing phenomenon, observable in the span of a human lifetime. The bacteria in Lenski's flasks and the fruit flies in experimental chambers are living testaments to the power of natural selection, mutation, and genetic drift. They reveal a world in constant flux, shaped by forces both random and predictable, where life persists and adapts through an unending dance of change.

In studying these organisms, we are not merely observing their evolution; we are deepening our understanding of our own. The same processes that drive change in bacteria and fruit flies have shaped the history of Homo sapiens, from the development of bipedalism to the emergence of language and culture. These experiments, though conducted on microscopic and diminutive scales, offer a window into the universal principles that govern all life—a reminder that, in the grand tapestry of evolution, we are both participants and observers.

Kendir Ramiz

The Fossil Record: The Historical Development of Direct Evidence

The fossil record is one of the most compelling and tangible pieces of evidence for evolution. It is a window into the distant past, preserving traces of life from epochs long before Homo sapiens ever walked the Earth. These remnants—bones, shells, imprints, and even entire organisms trapped in amber—are not just relics of a bygone world; they are chapters in a story that spans billions of years, chronicling the gradual transformation of life. Yet, the story told by fossils is neither linear nor complete. Like pieces of a jigsaw puzzle, they provide glimpses of the whole, inviting scientists to reconstruct the narrative of evolution with each new discovery.

The systematic study of fossils began in earnest during the 18th and 19th centuries, a period when the foundations of modern geology and paleontology were being laid. Early fossil hunters, often amateurs driven by curiosity, unearthed bones and shells that defied easy explanation. These discoveries challenged prevailing

Evolution and Creation Different Questions, One Path

worldviews, particularly the notion that Earth's history was a few thousand years old and unchanging. Among the first to grapple with these implications was Georges Cuvier, a French naturalist whose studies of extinct species demonstrated that the Earth had been home to forms of life radically different from those known in his time.

Cuvier's work marked a turning point in scientific thought. By studying the anatomical features of fossils, he was able to identify patterns of extinction and succession, suggesting that life on Earth had undergone profound changes over time. However, Cuvier rejected the idea of evolution, attributing these changes to catastrophic events. It was not until the publication of Charles Darwin's On the Origin of Species in 1859 that fossils were placed within an evolutionary framework, providing a mechanism—natural selection—to explain the observed patterns.

Darwin himself recognized the importance of the fossil record, though he acknowledged its limitations. In his time, the record was incomplete, with many gaps that seemed to challenge his theory. Critics pointed to the absence of transitional forms—species that bridged the evolutionary gap between major groups—as evidence against evolution. Darwin countered by arguing that fossilization was a rare event, requiring specific conditions that were unlikely to occur frequently. Over time, he predicted, the discovery of more fossils would fill these gaps and strengthen the case for evolution.

In the decades that followed, Darwin's prediction proved remarkably prescient. The discovery of Archaeopteryx in 1861, just two years after the publication of his book, provided one of the first clear examples of a transitional form. This ancient creature, with its blend of reptilian and avian features—teeth, claws, and a long bony tail alongside feathered wings—demonstrated the evolutionary link between dinosaurs and birds. Archaeopteryx was more than a curiosity; it was a validation of Darwin's theory, offering direct evidence of the gradual transformation of one group into another.

The 20th century brought an explosion of fossil discoveries, many of which illuminated key moments in the history of life. In the early 1900s, the unearthing of Homo erectus fossils in Java and Peking provided the first substantial evidence of human evolution, bridging the gap between modern humans and their ancient ancestors. These discoveries were followed by fossils of Australopithecus afarensis, such as the famous "Lucy" skeleton found in Ethiopia in 1974. Lucy, who lived approximately 3.2 million years ago, exhibited a combination of ape-like and human-like traits, underscoring the gradual evolution of bipedalism.

The mid-20th century also saw the discovery of entire ecosystems preserved in exceptional detail, such as the Burgess Shale in Canada. This fossil bed, dating back over 500 million years, contains an astonishing array of soft-bodied organisms from the Cambrian explosion, a period of rapid diversification in life forms. These fossils revealed a level of complexity and diversity that

Evolution and Creation Different Questions, One Path

challenged previous assumptions about early life, highlighting the evolutionary innovations that laid the foundation for modern ecosystems.

More recently, the study of fossils has been revolutionized by advances in technology. Techniques such as radiometric dating have allowed scientists to determine the age of fossils with remarkable precision, anchoring them within a detailed geological timeline. Meanwhile, tools like CT scanning and 3D modeling have provided new ways to analyze fossilized remains, uncovering details that were once hidden. For example, scans of dinosaur fossils have revealed the presence of soft tissues and pigments, offering insights into their physiology and even their coloration.

One of the most significant developments in paleontology has been the identification of transitional fossils that illustrate key stages in the evolution of major groups. For instance, the discovery of Tiktaalik in 2004 filled a critical gap in the transition from fish to tetrapods, the first vertebrates to venture onto land. Tiktaalik possessed features of both groups: fins with skeletal structures resembling limbs, a neck that allowed for greater mobility, and lungs adapted for breathing air. These traits demonstrated the gradual adaptations that enabled life to colonize terrestrial environments.

The fossil record is not without its challenges. Gaps remain, and the process of interpreting fossils requires careful analysis and, at times, revision. New discoveries can upend established narratives, as seen in debates over the relationships between hominin species or the causes of mass extinctions. Yet, these challenges are not weaknesses; they are a testament to the dynamic and self-correcting nature of science. Each fossil discovery adds a new piece to the puzzle, refining and enriching our understanding of life's history.

Perhaps the most profound lesson of the fossil record is its reminder of life's resilience and adaptability. It chronicles not only the emergence of new forms but also the losses—mass extinctions that wiped out entire lineages, reshaping the trajectory of evolution. These events, though catastrophic, paved the way for new opportunities, demonstrating the capacity of life to rebound and diversify in the face of change.

The study of fossils is more than a scientific pursuit; it is a meditation on time and transformation. Each fossil, from the tiniest trilobite to the towering skeletons of dinosaurs, is a fragment of a story that began billions of years ago and continues to this day. It is a story that connects all living things, reminding us that we are part of a continuum—a single thread in the vast tapestry of life. As we unearth these remnants of the past, we do more than reconstruct the history of evolution; we deepen our understanding of the forces that shaped us and our place within the ever-changing web of existence.

Evolution and Creation Different Questions, One Path

Genetic Relics: The Palmaris Longus and the Vestigial Third Eyelid

Evolution is a story written not only in fossils and ecosystems but also in the very fabric of our bodies. Hidden within our anatomy and genetic code are remnants of a distant past, silent witnesses to the journey of our species and the countless adaptations that shaped it. These evolutionary traces, known as vestigial structures, are like echoes of ancient worlds—once functional features that have become redundant or repurposed over time. Among the most fascinating of these relics are the palmaris longus, a muscle in the human arm, and the vestigial third eyelid, small and nearly forgotten yet rich with evolutionary significance.

The palmaris longus is a slender, ribbon-like muscle that runs from the elbow to the wrist. In many animals, particularly those adept at climbing or gripping, this muscle plays a critical role in enhancing grip strength. In primates like monkeys and apes, the palmaris longus supports their ability to swing through trees or hold onto branches—a survival skill of paramount importance. In humans, however, the muscle has lost much of its function. Approximately 10-15% of people are born without it entirely, and its absence has no noticeable impact on hand strength or dexterity.

Why, then, does the palmaris longus persist in the majority of the population? Evolution is rarely in a hurry to discard structures unless their presence imposes a significant cost. The palmaris longus, though functionally redundant in humans, does not interfere with our survival or reproduction. As a result, it remains as a benign artifact, a genetic memory of our arboreal ancestors. Surgeons, however, find a modern use for it. When reconstructing tendons in the hand or wrist, they often harvest the palmaris longus, repurposing this evolutionary leftover for medical innovation—a fitting example of adaptation within adaptation.

Another vestigial feature that links us to our evolutionary past is the plica semilunaris, a small fold of tissue located in the inner corner of the human eye. This structure is all that remains of a third eyelid, or nictitating membrane, a feature still functional in many animals, including birds, reptiles, and amphibians. In these species, the third eyelid serves as an additional layer of

Evolution and Creation Different Questions, One Path

protection, sliding across the eye to shield it from debris, maintain moisture, or assist in hunting underwater.

In humans, the third eyelid has atrophied to near invisibility, its protective function rendered obsolete by the evolutionary shift to more complex behaviors and environments. For our ancestors, whose lives depended less on swimming or hunting at high speeds and more on tool use and social cooperation, the nictitating membrane became unnecessary. Over generations, it diminished into the tiny crescent we see today, a vestige of a feature that once played a vital role.

Vestigial structures like the palmaris longus and the third eyelid are more than biological curiosities; they are compelling evidence for evolution. They demonstrate how traits are shaped by natural selection, retained when useful, and diminished when no longer necessary. The gradual fading of these features is not a sign of imperfection but of efficiency—a process by which life repurposes, adapts, and evolves in response to changing needs.

What makes these structures particularly fascinating is their ability to reveal our connections to other species. The palmaris longus links us to our primate relatives, while the third eyelid ties us to an even broader lineage, stretching back to the early vertebrates. These features remind us that Homo sapiens is not a singular creation but part of a vast and interconnected web of life. Every muscle, every fold of tissue carries a story—a record of the challenges and environments that shaped our ancestors.

But vestigial structures are not confined to the external anatomy. They are encoded within our genes, where dormant instructions for ancient traits occasionally resurface. For example, certain rare genetic conditions can result in humans being born with a tail, a feature that once played a critical role in balance and mobility for our primate ancestors. Though usually removed surgically in infancy, these tails serve as stark reminders of our evolutionary journey—a glimpse into the traits that were once essential but have since been left behind.

The persistence of such features also raises intriguing questions about the future of evolution. Will these vestiges continue to fade, or will they take on new roles? In a world where medical and technological advancements mitigate the pressures of natural selection, traits like the palmaris longus may persist indefinitely, carried along by the currents of genetic drift. Alternatively, they might be co-opted for new purposes, illustrating evolution's tendency to innovate from existing materials rather than start anew.

Vestigial structures also provoke reflection on what it means to be human. They remind us that our bodies are not static, perfect machines but dynamic, ever-changing systems shaped by millions of years of adaptation. They are a testament to the resilience and ingenuity of life, capable of surviving and thriving despite, or perhaps because of, these remnants of the past.

Evolution and Creation Different Questions, One Path

The palmaris longus and the vestigial third eyelid are small, almost trivial features, easily overlooked in the complexity of the human body. Yet, their significance lies not in their size or function but in the stories they tell. They are markers of a shared evolutionary history, connecting us to a world that existed long before our species emerged. They are symbols of change, evidence that life is not a fixed state but a process—a journey marked by both progress and imperfection.

In studying these genetic relics, we are reminded that evolution is not a distant, abstract phenomenon but an ongoing reality, inscribed within us. It is a process that has shaped every living being on this planet, from the bacteria in the soil to the humans reading these words. And it is a process that continues, quietly but inexorably, shaping the future in ways we can only begin to imagine. Through the lens of these small, unassuming structures, we catch a glimpse of the vast and intricate narrative of life—a story still unfolding, with us as both its products and its participants.

Kendir Ramiz

Molecular Evolution: The Role of DNA and Genetic Mutations

Beneath the surface of every living organism lies a molecular narrative, encoded in strands of DNA, that chronicles the journey of life. Long before the fossil record or vestigial structures could provide evidence of evolution, this genetic blueprint was quietly shaping the diversity and adaptability of species. DNA, the molecule of heredity, is not just a record of our biological history; it is an active participant in the evolutionary process. It mutates, recombines, and interacts with the environment, driving the changes that allow life to persist and thrive in an ever-changing world.

At the heart of molecular evolution is the mutation, a small yet profound alteration in the genetic code. Mutations can occur for various reasons—errors during DNA replication, exposure to radiation or chemicals, or the insertion of viral genetic material. Most mutations are neutral, having little to no impact on the organism. Some are deleterious, leading to diseases or reduced fitness. But a rare few are beneficial, conferring

Evolution and Creation   Different Questions, One Path

advantages that increase the likelihood of survival and reproduction. These beneficial mutations are the raw material of evolution, providing the variation upon which natural selection acts.

Consider the classic example of antibiotic resistance in bacteria. When exposed to antibiotics, most bacteria die, but a few may carry mutations that allow them to survive. These resistant individuals reproduce, passing their advantageous genes to the next generation. Over time, the population shifts, and the once-rare mutation becomes widespread. This process, observed in laboratories and hospitals alike, is evolution in action, driven by the molecular dynamics of DNA.

The study of molecular evolution has been revolutionized by advances in genomics, the large-scale analysis of DNA sequences. By comparing the genomes of different species, scientists can trace evolutionary relationships and identify the genetic changes that underlie key adaptations. For example, the genetic similarity between humans and chimpanzees—approximately 98.8% identical—highlights our shared ancestry while also pointing to the mutations that distinguish us as a species. These differences, though small in number, have profound effects on traits such as brain size, language ability, and bipedalism.

One of the most fascinating aspects of molecular evolution is the concept of gene duplication, a process that creates copies of existing genes. These duplicate genes can acquire new functions through mutation, leading to evolutionary innovation. The evolution of hemoglobin, the protein responsible for oxygen transport in the blood, is a striking example. Gene duplication events allowed the original oxygen-carrying protein to diversify into specialized forms, such as fetal hemoglobin, which binds oxygen more effectively in the low-oxygen environment of the womb. This molecular adaptation has had profound implications for mammalian development and survival.

Another powerful tool in studying molecular evolution is the molecular clock, a method that uses the rate of genetic mutations to estimate the timing of evolutionary events. By calibrating this clock with known fossil dates, scientists can reconstruct the timeline of life's history with remarkable precision. For instance, molecular clock studies have revealed that humans and chimpanzees diverged from a common ancestor approximately 6-7 million years ago—a timeline consistent with fossil evidence.

Molecular evolution also sheds light on the interplay between genetic drift and natural selection. While selection favors advantageous mutations, drift introduces randomness, particularly in small populations. This randomness can fix neutral or even slightly deleterious mutations, shaping the genetic landscape in ways that are not solely driven by fitness. The balance between these forces underscores the complexity of evolution, where chance and necessity intertwine to produce the diversity of life.

Evolution and Creation Different Questions, One Path

One of the most intriguing discoveries in molecular evolution is the presence of conserved genes, sequences that have remained largely unchanged over vast stretches of time. These genes, such as those involved in basic cellular processes like DNA replication and energy metabolism, highlight the shared molecular foundation of life. The fact that a gene in yeast can perform the same function in humans speaks to the deep evolutionary connections that unite all organisms. These conserved elements are a reminder that even in the face of immense diversity, life is built on a common molecular framework.

But not all genetic changes are beneficial or neutral. Some mutations lead to genetic disorders, highlighting the trade-offs inherent in evolution. For example, the mutation that causes sickle cell anemia is harmful in individuals who inherit two copies of the gene. Yet, in regions where malaria is prevalent, carrying a single copy provides resistance to the disease, illustrating how selection can maintain harmful alleles under certain conditions. This delicate balance between benefit and cost is a recurring theme in molecular evolution, shaping the trajectories of species in ways that are both subtle and profound.

Molecular evolution also invites us to consider the future of our species. Advances in genetic engineering and synthetic biology are giving humans unprecedented control over their own DNA, raising questions about the nature of evolution in the age of technology. Will we continue to evolve naturally, or will we take the reins, directing our own genetic destiny? The potential to edit genes, eradicate diseases, or enhance traits offers both promise and peril, challenging us to navigate the intersection of biology and ethics with care.

The study of DNA and genetic mutations reveals evolution not as a theoretical construct but as a living, ongoing process. It connects the microcosm of molecular change to the macrocosm of species diversity, linking the tiniest nucleotide substitution to the grand patterns of life's history. Each mutation, each duplication, each recombination is a step in a journey that has been unfolding for billions of years—a journey that continues, even now, within the cells of every living organism.

In exploring molecular evolution, we uncover not only the mechanisms of change but also the profound interconnectedness of life. The genetic code, shared by all organisms, is a testament to our common origins and a reminder of the continuity that binds us to every creature on this planet. It is a story written in the language of DNA, a narrative of adaptation, innovation, and resilience that stretches from the dawn of life to the present moment. And it is a story that continues to unfold, offering new insights with every discovery, challenging us to see ourselves as both participants in and stewards of this extraordinary evolutionary legacy.

Evolution and Creation Different Questions, One Path

Antibiotic Resistance: Evolution's Impact on Everyday Life

Evolution is often perceived as a slow and distant process, unfolding over millennia and observable only in fossils or ancient genetic markers. Yet, evolution is not confined to the distant past. It is a dynamic and ongoing phenomenon, shaping the world around us in ways that are both subtle and profound. One of the most striking examples of evolution in action can be found in the rise of antibiotic resistance, a phenomenon that touches our everyday lives and underscores the immediacy of evolutionary principles.

Kendir Ramiz

Antibiotics, hailed as one of the greatest medical breakthroughs of the 20th century, revolutionized healthcare by providing effective treatments for bacterial infections. Diseases that once claimed millions of lives—such as tuberculosis, pneumonia, and syphilis—were brought under control, saving countless lives and reshaping human history. Yet, even as antibiotics transformed medicine, they also set the stage for a profound evolutionary struggle, one in which bacteria, through the mechanisms of mutation and natural selection, began to adapt and resist.

At the heart of antibiotic resistance lies the principle of selective pressure, a cornerstone of evolutionary theory. When antibiotics are introduced into an environment, they create a hostile landscape for bacteria. The majority of bacterial cells, lacking resistance mechanisms, are killed off. However, in any given population, there are often a few individuals that possess random genetic mutations enabling them to survive. These mutations might allow the bacteria to break down the antibiotic, pump it out of their cells, or modify the target that the antibiotic attacks.

These resistant bacteria, now free from competition, proliferate rapidly, passing their advantageous mutations to subsequent generations. Over time, the population shifts, becoming dominated by resistant strains. This is not just evolution in action—it is evolution happening at a speed that highlights the adaptability and resilience of life. In the span of mere decades, bacteria have transformed from vulnerable pathogens into formidable adversaries, challenging the very foundations of modern medicine.

One of the most well-documented examples of antibiotic resistance is methicillin-resistant Staphylococcus aureus (MRSA), a strain of bacteria that has become resistant to multiple classes of antibiotics. MRSA infections, once confined to hospitals, have spread into the community, causing severe and sometimes fatal illnesses. Similarly, drug-resistant strains of tuberculosis, gonorrhea, and E. coli have emerged, complicating treatment and raising the specter of a post-antibiotic era in which common infections become deadly once again.

The rapid rise of antibiotic resistance is not solely a result of bacterial ingenuity; it is also a consequence of human behavior. The overuse and misuse of antibiotics in medicine, agriculture, and even household products have accelerated the process. Antibiotics are often prescribed unnecessarily for viral infections, such as the common cold, against which they are ineffective. In agriculture, antibiotics are routinely used to promote growth in livestock, creating reservoirs of resistant bacteria that can spread to humans through food and the environment. These practices create a perfect storm, amplifying the selective pressures that drive resistance.

The implications of antibiotic resistance extend far beyond individual infections. They threaten the entire ecosystem of modern medicine, which relies on effective antibiotics to perform surgeries, treat cancer patients, and manage chronic illnesses. Without antibiotics, the risk of infection becomes a barrier to medical progress, undoing decades of advancements and forcing a reevaluation of how we approach healthcare.

Despite the challenges, the rise of antibiotic resistance also offers valuable lessons about the power and predictability of evolution. It demonstrates the interconnectedness of life, showing how actions in one domain—such as the use of antibiotics in agriculture—can ripple across ecosystems and affect human health. It also highlights the necessity of adapting our strategies to align with the realities of evolutionary dynamics.

Addressing antibiotic resistance requires a multifaceted approach rooted in both science and societal change. On a practical level, efforts are underway to develop new antibiotics that can overcome resistance, though the process is costly and time-consuming. At the same time, researchers are exploring alternative therapies, such as phage therapy, which uses viruses that specifically target bacteria, or the development of drugs that disable resistance mechanisms rather than kill bacteria outright.

Equally important is the need for responsible antibiotic stewardship. This includes reducing the unnecessary use of antibiotics, both in healthcare settings and in agriculture, and educating the public about the dangers of misuse. It also involves improving infection prevention measures, such as vaccination, sanitation, and hygiene, to reduce the need for antibiotics in the first place. These efforts must be global, reflecting the reality that antibiotic resistance knows no borders and that bacteria, unlike humans, do not recognize national boundaries.

The story of antibiotic resistance is not merely a cautionary tale; it is a living example of evolution's relevance to everyday life. It reminds us that evolution is not an abstract or historical process but a force that continues to shape the world around us. It also challenges us to think critically about our relationship with the natural world, to recognize the unintended consequences of our actions, and to embrace the principles of adaptability and resilience that evolution teaches.

In confronting antibiotic resistance, we are not only addressing a medical crisis but also engaging with one of the fundamental dynamics of life: the constant interplay between change and survival. It is a reminder that evolution, while often invisible, is ever-present, shaping the trajectory of life in ways both subtle and profound. The choices we make today will determine not only the future of medicine but also the legacy of our species within the broader story of life on Earth.

Kendir Ramiz

The Shared Language of Science and Faith: The Harmony of Creative Design and Scientific Laws

Throughout history, science and faith have often been cast as adversaries, locked in a perpetual struggle for the soul of human understanding. Science seeks to uncover the mechanisms of the universe through observation and experimentation, while faith often offers a sense of purpose and meaning, grounded in the belief in a higher power or creative force. On the surface, these domains might appear irreconcilable, with science rooted in empirical evidence and faith anchored in transcendental truths. Yet, beneath this apparent tension lies a profound harmony—a shared language that speaks to the intricate balance of the universe and the human desire to comprehend it.

To understand this harmony, one must first consider the nature of scientific laws. These laws, whether describing gravity, thermodynamics, or evolution, are not arbitrary constructs but reflections of the consistent patterns that govern the cosmos. They reveal a universe that is both ordered and intelligible, one where phenomena can be

Evolution and Creation Different Questions, One Path

predicted and understood through reason and inquiry. The very existence of such order—why the universe operates according to laws at all—is a question that has long intrigued scientists and theologians alike. For many, it suggests not a random chaos but an underlying design, a creative force that imbued the cosmos with structure and coherence.

This idea resonates deeply with many religious traditions, which describe a universe brought into existence through intentionality. In the Abrahamic faiths, the act of creation is described as the word of God setting the cosmos into motion—a divine utterance that brought order from chaos. In Hinduism, the concept of Rta embodies a similar notion: the cosmic order that sustains the balance of the universe. These perspectives do not reject the idea of natural laws but embrace them as expressions of a deeper, transcendent truth.

The intersection of science and faith is perhaps most evident in the study of evolution, a field that has often been a flashpoint for controversy. At its core, evolution is a process governed by natural laws—mutation, selection, and adaptation—that operate consistently over time. For those who view the universe as the work of a creator, these laws can be seen not as contradictions to divine intention but as the tools through which creation unfolds. The process of evolution, with its capacity for innovation and complexity, mirrors the creativity often attributed to a divine force. It suggests a world where life is not static but dynamic, capable of change and growth in ways that reflect the vastness of the cosmos itself.

One of the most compelling attempts to bridge science and faith comes from the field of theistic evolution, which posits that evolutionary processes are not random but guided by a divine plan. Proponents like Francis Collins, the renowned geneticist and leader of the Human Genome Project, have argued that the laws of nature are evidence of a creator who established the parameters of the universe and allowed life to emerge within them. Collins, a devout Christian, describes the genetic code as "the language of God," a testament to the elegance and universality of the processes that govern life. For him, and many others, the discoveries of science do not diminish faith but deepen it, revealing a universe imbued with purpose and meaning.

The idea that science and faith share a common language extends beyond biology to the very fabric of the universe. Consider the fine-tuning of the physical constants that govern reality—quantities like the speed of light, the gravitational constant, and the strength of the electromagnetic force. If these values were even slightly different, the universe as we know it could not exist. Stars would not form, chemical reactions would not occur, and life would be impossible. For many scientists, this fine-tuning is a profound mystery, one that has led some to speculate about the existence of a multiverse—a vast ensemble of universes, each with its own set of constants. For others, it suggests

Evolution and Creation Different Questions, One Path

intentionality, a creative force that calibrated the universe with exquisite precision.

But the harmony of science and faith is not limited to grand cosmological questions; it also finds expression in the human experience. Both domains seek to address the fundamental questions of existence: Where do we come from? Why are we here? What is our place in the cosmos? While science provides answers rooted in evidence and logic, faith offers a framework for meaning and purpose, guiding how we interpret and engage with the world. Together, they form a complementary lens through which humanity can explore the mysteries of existence.

This harmony is perhaps best understood as a dialogue rather than a synthesis. Science and faith are not identical, nor do they need to be. Each has its own methods, goals, and limitations. Science excels at explaining the "how" of phenomena—how galaxies form, how DNA replicates, how evolution shapes life. Faith, on the other hand, often addresses the "why"—why the universe exists, why life has meaning, why humans possess the capacity for love and morality. When these domains are allowed to inform and enrich one another, they offer a more holistic understanding of the universe and our place within it.

This dialogue also invites a sense of humility. Science, for all its achievements, acknowledges the limits of human knowledge. There are questions that remain unanswered, mysteries that defy explanation. Faith, too, recognizes the vastness of the divine, often describing a creator beyond human comprehension. Together, they remind us that the pursuit of understanding is not about certainty but about exploration, a journey that is as much about the questions as it is about the answers.

The shared language of science and faith is not a call for uniformity but for unity—a recognition that the quest to understand the universe is a deeply human endeavor, one that transcends disciplines and beliefs. It is a reminder that the cosmos, in all its complexity and wonder, is a gift—whether viewed as the product of a creator, the outcome of natural processes, or both. This perspective does not diminish the rigor of science or the depth of faith; instead, it elevates both, revealing the interconnectedness of all knowledge and the shared curiosity that drives humanity forward.

In exploring the harmony of creative design and scientific laws, we are reminded of our place within the universal order—not as masters or observers but as participants in a story that is still unfolding. The laws of nature, the insights of science, and the mysteries of faith all converge to paint a picture of a universe that is both knowable and wondrous, a testament to the enduring power of human curiosity and the profound beauty of existence itself.

Evolution and Creation Different Questions, One Path

From Myth to Reality: Celestial Civilizations and Creation Myths

Myths are among humanity's earliest attempts to make sense of the universe. Long before telescopes unveiled distant galaxies or genetics unraveled the code of life, ancient cultures told stories to explain the origins of existence, the nature of humanity, and the forces that govern the cosmos. These myths were not merely fanciful tales; they were frameworks for understanding, deeply interwoven with the social, spiritual, and intellectual fabric of their time. Among the most intriguing of these myths are those that speak of celestial civilizations—beings from the heavens who shaped the Earth and humanity itself.

Kendir Ramiz

From the Sumerians to the Mayans, nearly every ancient culture carries echoes of creation stories involving divine or otherworldly entities. These narratives often describe powerful beings descending from the sky to create, guide, or rule humankind. In the Sumerian tradition, the Anunnaki, deities believed to originate from the heavens, were said to have played a direct role in shaping human civilization. They were depicted not only as creators but also as administrators, establishing systems of order and governance. These tales, rich with symbolism, offer a glimpse into how early humans understood their relationship to the cosmos and to forces beyond their control.

Similarly, Hindu mythology speaks of celestial beings, such as the Devas, who are portrayed as stewards of natural forces, balancing creation and destruction in a cosmic dance. The Vedas, some of humanity's oldest sacred texts, weave intricate stories of these celestial entities, blending philosophical musings with vivid imagery. The Mayans, too, envisioned a world shaped by otherworldly forces, with their gods descending to Earth to create humans out of maize—a substance considered sacred and central to their survival.

These myths are not isolated phenomena but part of a global pattern. Across continents and cultures, we find recurring themes: beings from the heavens interacting with Earth, imparting knowledge, and shaping human destiny. The striking similarities between these stories raise questions that continue to captivate scholars,

Evolution and Creation Different Questions, One Path

scientists, and storytellers alike. Are these myths purely symbolic, or do they hint at something more tangible—an ancient understanding of forces or beings beyond human comprehension?

One interpretation, particularly popularized in modern times, is the idea of ancient astronauts—advanced extraterrestrial civilizations visiting Earth in the distant past. Proponents of this theory point to mythological accounts as potential evidence, suggesting that descriptions of gods and celestial beings may be rooted in real encounters with technologically advanced visitors. They highlight artifacts, architectural feats, and unexplained technological advancements as supporting evidence. For instance, the precise alignment of the Great Pyramid of Giza with celestial bodies, or the detailed astronomical knowledge embedded in Mayan and Mesopotamian records, are often cited as indicators of otherworldly influence.

While such theories ignite the imagination, they also draw criticism for oversimplifying and misinterpreting complex cultural legacies. Skeptics argue that humanity's ingenuity and resourcefulness are sufficient to explain these achievements, without invoking extraterrestrial intervention. They emphasize the importance of viewing myths within their cultural and historical contexts, where celestial beings often symbolize natural phenomena, moral principles, or philosophical ideas rather than literal visitors from the stars.

But regardless of whether these myths describe real encounters or symbolic narratives, they reveal something profound about the human psyche: an enduring fascination with the heavens and a desire to connect with something greater than ourselves. Ancient people gazed at the stars with wonder, crafting stories to bridge the gap between the known and the unknown. These stories, passed down through generations, served as both explanations of natural phenomena and as guides for navigating life's mysteries.

Modern science has not diminished this sense of wonder. On the contrary, our exploration of the cosmos has reignited it. The discovery of exoplanets, the study of extremophiles—organisms that thrive in conditions once thought uninhabitable—and the ongoing search for extraterrestrial intelligence (SETI) have brought the possibility of celestial civilizations from the realm of myth into the realm of scientific inquiry. What ancient myths described poetically, modern science approaches empirically, asking questions about the origins of life, the conditions necessary for its development, and the likelihood of intelligent beings elsewhere in the universe.

The intersection of myth and science offers a unique lens through which to explore these questions. For instance, the idea of panspermia, which suggests that life may have originated elsewhere in the universe and been transported to Earth via comets or meteors, echoes mythological themes of life being seeded by celestial forces. Similarly, the search for extraterrestrial civilizations—whether through radio signals or telescopic surveys—mirrors humanity's ancient longing to connect with the heavens.

Yet, myths do more than speculate about the origins of life; they also reflect humanity's aspirations, fears, and values. The stories of celestial beings descending to Earth often involve the transfer of knowledge or technology, a theme that resonates with our own experience of progress. These myths suggest a hope that contact with the divine—or the extraterrestrial—might elevate humanity, offering guidance or wisdom to navigate an uncertain world.

In reexamining these myths through the lens of modern understanding, we are reminded of the continuity between ancient and contemporary thought. The questions posed by our ancestors—about the nature of the cosmos, the origins of life, and humanity's place in the universe—are the same questions that drive our scientific endeavors today. Myths, far from being relics of a bygone era, remain relevant as expressions of humanity's unending curiosity and imagination.

As we venture further into the cosmos, sending probes to distant planets and listening for signals from the stars, we are not leaving behind the myths of the past but building upon them. They serve as a foundation for our exploration, a reminder that the quest to understand the universe is as old as humanity itself. Whether we find evidence of celestial civilizations or not, the act of searching connects us to a legacy of wonder and inquiry that transcends time and space.

Kendir Ramiz

The transition from myth to reality is not a dismissal of the former but an evolution of perspective. Myths capture truths that science often struggles to articulate—truths about meaning, purpose, and the human connection to the cosmos. As we explore the universe, we carry these truths with us, blending the poetic and the empirical in a journey that is both ancient and ongoing. Through this interplay, we discover not only the secrets of the stars but also the enduring nature of our own humanity.

Evolution and Creation Different Questions, One Path

The Gaps in Homo Sapiens' Evolution

The story of Homo sapiens is one of astonishing transformation. From humble beginnings as a small population of primates on the African savanna, our species has risen to dominate the planet, shaping the environment, exploring the cosmos, and even pondering its own origins. Yet, for all our achievements, the evolutionary journey that brought us here remains riddled with mysteries. Beneath the surface of what we think we know lies a series of unanswered questions and unexplained leaps—gaps in the record that challenge our understanding of who we are and how we came to be.

To begin with, the origins of Homo sapiens are remarkably clear in some respects and deeply enigmatic in others. Fossil evidence and genetic studies point to a common ancestor shared with Neanderthals and Denisovans, tracing back roughly 600,000 years. From this lineage, Homo sapiens emerged around 300,000 years ago, with early fossils discovered in Morocco offering a glimpse of anatomically modern humans. Yet, these discoveries also raise questions. The Moroccan fossils, for instance, exhibit a curious mix of modern and archaic traits, suggesting a complex web of evolutionary pathways rather than a single linear progression.

One of the most puzzling gaps in our evolutionary narrative lies in the sudden cognitive leap that distinguishes Homo sapiens from other hominin species. While our ancestors shared physical traits with Neanderthals and Denisovans—such as similar brain sizes and robust physiques—modern humans exhibit an extraordinary capacity for abstract thought, symbolic language, and complex social structures. The emergence of art, ritual, and technology around 70,000 years ago, often referred to as the Cognitive Revolution, marks a turning point in human evolution. But what triggered this transformation? Why did Homo sapiens, alone among the hominins, develop these capabilities?

Theories abound, ranging from genetic mutations to cultural innovation. Some researchers point to changes in brain structure or connectivity, such as the expansion of the prefrontal cortex or increased neural plasticity. Others suggest that a mutation in the FOXP2 gene,

Evolution and Creation Different Questions, One Path

associated with language development, may have given our species a communicative edge. Still others argue that environmental pressures—such as dramatic climate fluctuations during the Pleistocene—forced early humans to adapt in novel ways, fostering creativity and collaboration.

Yet, these explanations, while compelling, remain incomplete. They cannot fully account for the speed and scope of the changes observed in Homo sapiens. How did these traits evolve so rapidly, and why did they not arise in other hominins with similar brains and environments? The gaps in our understanding invite not only scientific inquiry but also philosophical reflection, challenging us to reconsider the uniqueness of our species and the forces that shape evolution.

Another enigma lies in the fate of our closest relatives. For tens of thousands of years, Homo sapiens shared the Earth with other hominins, including Neanderthals in Europe, Denisovans in Asia, and possibly other yet-to-be-identified species. These groups interbred to some extent, leaving traces of their DNA in modern human populations. Yet, by around 30,000 years ago, all other hominins had vanished, leaving Homo sapiens as the sole survivor of a once-diverse lineage. What caused their extinction? Was it competition with Homo sapiens, environmental change, or some combination of factors?

The answers are elusive, but recent discoveries offer intriguing clues. Fossils and archaeological sites suggest that Neanderthals and Denisovans were not the brutish, unsophisticated creatures once imagined but intelligent beings capable of tool use, art, and even ritual. Their disappearance may not have been the result of direct conflict with Homo sapiens but rather the gradual erosion of their populations through competition for resources, environmental pressures, and genetic assimilation. Still, the precise dynamics of these interactions remain a subject of debate, underscoring the complexity of human evolution.

Even within our own species, the evolutionary record is marked by discontinuities and surprises. The transition from small, scattered hunter-gatherer groups to large, sedentary civilizations is one of the most dramatic shifts in human history. Yet, the mechanisms that enabled this shift are not fully understood. Why did Homo sapiens develop agriculture and complex societies while other species did not? Was it a matter of chance, necessity, or some intrinsic characteristic of our species? The answers may lie not only in biology but also in the interplay of culture, environment, and innovation—a multifaceted process that defies simplistic explanations.

Perhaps the most profound mystery of all is the role of chance in our evolution. Evolution is often portrayed as a deterministic process, governed by natural selection and adaptive pressures. But the history of life is also shaped by random events, from genetic mutations to asteroid impacts. Was the rise of Homo sapiens an inevitable outcome of evolutionary forces, or the result of a series of lucky breaks? Would intelligent life have emerged on Earth if conditions had been slightly different, or are we an anomaly in the vast expanse of the cosmos?

These questions, far from diminishing the significance of our species, highlight the intricate web of factors that shape evolution. They remind us that our understanding of human origins is a work in progress, a mosaic of evidence that continues to evolve with each new discovery. Fossils, genes, and tools are not just artifacts of the past; they are pieces of a puzzle that connects us to a lineage stretching back millions of years—a lineage filled with triumphs, failures, and unanswered questions.

In pursuing the unknown, we are not merely seeking to fill gaps in the record; we are engaging with the essence of what it means to be human. The search for our origins is both a scientific endeavor and a deeply existential one, driven by the same curiosity and imagination that propelled our ancestors to explore new horizons. It is a quest that unites us across time and space, linking the distant past to the unfolding future.

The gaps in Homo sapiens' evolution are not voids to be feared but opportunities to expand our understanding. They invite us to question, to explore, and to embrace the uncertainty that lies at the heart of discovery. In the uncharted territories of our evolutionary history, we find not only the story of how we came to be but also the inspiration to continue the journey—a journey that, like evolution itself, is ever unfinished and filled with possibility.

Kendir Ramiz

The Endless Journey: The Unfinished Story of Evolution

Evolution is not a tale with a clear beginning or a definitive end. It is not a linear progression toward perfection, nor a narrative with predictable chapters and characters. Rather, it is a ceaseless, sprawling journey—a dynamic process that unfolds across billions of years, shaped by chance and necessity, by adaptation and extinction, and by the relentless drive of life to persist and transform. To speak of evolution is to tell a story that is still being written, its final pages as unknowable as the first spark of life that ignited it.

Evolution and Creation Different Questions, One Path

From the perspective of deep time, our species, Homo sapiens, occupies but a fleeting moment in the grand chronology of evolution. The Earth itself, a cradle of unimaginable biodiversity, has existed for over 4.5 billion years, yet humans emerged only in the last 300,000 years—a blink in the geological eye. Before us came epochs dominated by trilobites, dinosaurs, and megafauna, each leaving their mark before fading into the annals of extinction. Yet, these disappearances were not endpoints but transitions, opening new opportunities for life to diversify and adapt. Evolution, in this sense, is less a series of endings and beginnings and more a continuum—a journey of constant renewal.

For all its continuity, evolution is far from predictable. It is marked by punctuated equilibria, periods of stability disrupted by sudden bursts of change. These shifts can be triggered by environmental upheavals, mass extinctions, or the emergence of novel traits. The extinction of the dinosaurs 66 million years ago, brought about by a catastrophic asteroid impact, is a poignant example. This single event, devastating as it was, cleared ecological niches that allowed mammals—and eventually humans—to flourish. In this way, evolution demonstrates an uncanny resilience, transforming destruction into opportunity, chaos into creation.

But the story of evolution is not confined to the distant past; it is alive in the present, shaping the world around us and within us. Every living organism, from the simplest bacterium to the most complex human, is a product of evolutionary processes that continue to operate. Even in the span of a single lifetime, evolution can be observed in the development of antibiotic resistance in bacteria, the adaptation of species to urban environments, or the rapid spread of genetic traits in isolated populations. These changes remind us that evolution is not a relic of ancient history but an ongoing force that governs life in all its forms.

For Homo sapiens, evolution is both a biological and cultural phenomenon. While our genes still respond to natural selection, our ability to shape our environment has introduced new dynamics into the evolutionary equation. Agriculture, medicine, and technology have altered the selective pressures we face, enabling us to thrive in conditions that would have been insurmountable for our ancestors. Yet, these advancements come with their own challenges. As we modify our world, we must also confront the consequences of our actions—climate change, habitat destruction, and the erosion of biodiversity. In doing so, we are reminded that we are not separate from the evolutionary story but active participants in it, with the power to shape its trajectory.

The future of evolution raises profound questions about the nature of life and humanity's place within it. Will Homo sapiens continue to evolve biologically, or have we reached a plateau, our adaptations now driven primarily by culture and technology? Will artificial intelligence and genetic engineering blur the boundaries between biology and innovation, creating new forms of life that challenge our understanding of what it means to

Evolution and Creation Different Questions, One Path

be human? And what of the countless other species with whom we share this planet—how will their fates intertwine with ours as the forces of evolution and human activity collide?

These questions underscore the open-ended nature of evolution. Unlike a story with a predetermined ending, evolution has no fixed destination. Its path is shaped by countless variables, from the mutation of a single gene to the collision of celestial bodies. This unpredictability is both its challenge and its beauty, a reminder that life's resilience lies in its capacity for change. Evolution thrives not because it is planned but because it is adaptable, capable of responding to the unforeseen with innovation and complexity.

In contemplating the endless journey of evolution, we are also invited to reflect on our own journey as a species. Homo sapiens is unique not only in its capacity for abstract thought but also in its ability to shape its destiny. Yet, this power comes with responsibility. The choices we make today—about how we treat our environment, how we interact with other species, and how we harness the tools of science and technology—will influence not only our own evolutionary path but that of life itself.

The story of evolution is, in many ways, a story of connection. It reminds us that we are not isolated beings but part of a vast, interconnected web of life that stretches across time and space. The genes we carry are echoes of ancient ancestors, shared with countless other organisms that inhabit this planet. The air we breathe, the water we drink, and the ecosystems we rely on are products of evolutionary processes that have shaped the Earth for billions of years. To understand evolution is to understand that our fate is inextricably linked to the fate of the planet—a truth both humbling and empowering.

The journey of evolution is not about reaching a final destination but about embracing the process itself. It is a journey of discovery, adaptation, and renewal, one that invites us to see the world not as static but as dynamic, not as finished but as ever-unfolding. In this journey, we find not only the story of life's past but also the possibilities of its future—a future that is ours to explore, shape, and sustain.

Evolution and Creation Different Questions, One Path

Forging Our Path: Humanity's Quest for Growth Within the Universal Order

Human beings are unique in their ability to question, reflect, and envision. Unlike other species, which adapt to their environments through the slow mechanisms of evolution, Homo sapiens has developed the capacity to reshape its world and itself. Yet, this power brings with it a profound challenge: How do we align our ambitions with the universal order that has given rise to us? How do we navigate a world governed by natural laws and interdependence while striving for individual and collective growth? The quest to forge our path as a species, while respecting the intricate balance of the cosmos, is both our greatest challenge and our defining endeavor.

Kendir Ramiz

At the heart of this journey lies the human drive for self-improvement. This drive is not merely about survival or reproduction, as evolutionary imperatives might suggest. It is about transcendence—the desire to surpass our limitations, to understand ourselves and the universe more deeply, and to leave a legacy that extends beyond our lifetimes. This impulse manifests in countless ways: the pursuit of knowledge, the creation of art, the development of technologies, and the quest for moral and spiritual growth. It is an impulse that has propelled humanity from caves to cities, from myths to science, and from the Earth to the stars.

Yet, this drive is also fraught with contradictions. In our pursuit of growth, we often act in ways that disrupt the very systems that sustain us. Industrialization, while elevating billions out of poverty, has also led to environmental degradation and climate change. Technological advancements, while expanding our horizons, have created new ethical dilemmas and unforeseen consequences. Even our individual quests for success and fulfillment can sometimes come at the expense of community and connection. These contradictions are not failures but reflections of the complexity of being human. They remind us that growth is not a straightforward path but a delicate balance between ambition and humility.

To navigate this balance, we must first understand our place within the universal order. Science has shown us that we are not separate from the natural world but deeply interconnected with it. The air we breathe, the

Evolution and Creation Different Questions, One Path

food we eat, and the ecosystems we depend on are products of processes that have unfolded over billions of years. Our bodies, composed of the same elements that make up stars, are reminders of our shared origins with the cosmos. This understanding invites a shift in perspective: from seeing ourselves as masters of the universe to participants in a vast and intricate system.

This shift does not mean abandoning our aspirations but grounding them in a deeper awareness of the consequences of our actions. It means recognizing that our growth as individuals and as a species must be aligned with the principles of sustainability and interdependence. This alignment is not a limitation but a source of strength, allowing us to build a future that honors both our potential and our responsibility.

One of the most powerful ways to forge this alignment is through the cultivation of self-awareness. Self-awareness is not just about understanding our thoughts and emotions; it is about recognizing the broader context of our existence. It is about asking questions that go beyond immediate concerns: What kind of world do we want to create? What values will guide our actions? How can we contribute to the well-being of others and the planet? These questions are not easy to answer, but they are essential for forging a path that is both meaningful and sustainable.

Education plays a crucial role in this process. An education that fosters curiosity, critical thinking, and empathy can empower individuals to navigate the complexities of the modern world. It can help us move beyond simplistic narratives and binary thinking, embracing the nuances and interconnections that define our reality. It can also inspire a sense of wonder and responsibility, reminding us that knowledge is not just a tool for progress but a bridge to understanding and harmony.

Another essential aspect of forging our path is the recognition of our shared humanity. In a world often divided by borders, ideologies, and identities, it is easy to lose sight of the commonalities that unite us. Yet, the challenges we face—climate change, pandemics, inequality—are global in scope and require collective action. These challenges remind us that our fates are intertwined, that the well-being of one is connected to the well-being of all. Embracing this interconnectedness is not just a moral imperative but a practical necessity, guiding us toward solutions that benefit both individuals and the collective.

Finally, forging our path requires a commitment to imagination and creativity. The problems of the 21st century cannot be solved with the same thinking that created them. They demand new ideas, new approaches, and a willingness to challenge the status quo. This creativity is not limited to science or technology; it extends to how we think about relationships, governance, and the meaning of life itself. It invites us to reimagine what it means to thrive as individuals and as a species, to dream of a future that is not just sustainable but flourishing.

Evolution and Creation Different Questions, One Path

The journey of forging our path is not one of perfection but of progress. It is a process of learning and unlearning, of striving and stumbling, of confronting our limitations and embracing our potential. It is a journey that requires courage and compassion, reason and intuition, discipline and flexibility. Above all, it is a journey that calls us to see ourselves not as isolated beings but as integral parts of a larger whole—a universe that is as dynamic and evolving as we are.

As we navigate this journey, we are reminded that the universal order is not a constraint but an invitation. It invites us to explore, to create, and to grow, not in defiance of its principles but in harmony with them. It invites us to see our aspirations not as separate from the cosmos but as expressions of its inherent dynamism and creativity. And it invites us to embrace the unknown, not with fear but with curiosity, knowing that the path we forge is both a continuation of the past and a step into the future.

In forging our path, we are not only shaping the destiny of Homo sapiens but contributing to the ongoing story of life itself—a story that is as vast as the universe and as intimate as the choices we make each day. This is the essence of our humanity: the ability to grow, to connect, and to find meaning within the boundless possibilities of existence.

Kendir Ramiz

The Thread That Connects Us

Our journey through evolution, creation, and humanity's place within the universal order has not been a quest for definitive answers but an exploration of questions that define us. Who are we? Where do we come from? Where are we headed? These are not new questions. They have been asked in the flickering firelight of ancient caves, inscribed in sacred texts, debated in lecture halls, and pondered in solitude. And yet, for all our advancements, they remain as alive and urgent today as they were millennia ago.

What has emerged from this exploration is not a singular narrative but a tapestry—a richly woven fabric of science and myth, fact and meaning, the measurable and the ineffable. Each thread in this tapestry, whether fossilized evidence or poetic myth, contributes to a larger understanding of what it means to be human. And at the heart of it all is a paradox: the more we uncover about the universe, the more we realize how much remains unknown. But this unknowing is not a flaw; it is an invitation—a call to continue questioning, seeking, and imagining.

Evolution and Creation Different Questions, One Path

Evolution teaches us that change is the only constant. Life, in all its forms, adapts and evolves, not toward perfection but toward persistence. And yet, Homo sapiens is unique in its ability to shape its evolution consciously. We are not bound to the natural world in the same way as our ancestors, but neither are we separate from it. This duality—the power to shape and the necessity to adapt—places us in a position of immense responsibility. What we do with this power will determine not only our future but the futures of countless other species and ecosystems intertwined with our own.

As we look to that future, we find ourselves at a crossroads. Technology, genetics, and artificial intelligence offer possibilities that our ancestors could scarcely have imagined. But these advancements also bring ethical challenges and unintended consequences. Will we use our tools to enhance life or to diminish it? Will we create a world of shared opportunity or deepened inequality? These are not questions for scientists or philosophers alone; they are questions for all of humanity, requiring collective reflection and action.

Yet, amid these uncertainties, there is also hope. Hope in the resilience of life, in the creativity of the human spirit, and in the capacity for connection that transcends time and space. The same curiosity that drove our ancestors to paint on cave walls drives us to explore the stars. The same drive for survival that allowed early humans to adapt to harsh climates now inspires us to tackle global challenges like climate change. And the same capacity for wonder that gave rise to myths and religions continues to guide us as we grapple with the mysteries of existence.

In this vast and interconnected universe, we are both infinitesimally small and profoundly significant. Our atoms were forged in the hearts of stars, and our actions ripple outward in ways we cannot always predict. We are threads in a story that began billions of years ago, one that stretches beyond our comprehension yet invites our participation. To be human is to embrace this tension—to recognize our limits while reaching for the infinite.

As we close this chapter, let us not see it as an ending but as a pause in an ongoing dialogue. The questions we have explored are not meant to be resolved but to spark further inquiry. They remind us that knowledge is not a destination but a journey, one that requires humility, imagination, and an unwavering commitment to understanding.

We are creatures of evolution, but we are also creators of meaning. And in this dual role, we find our purpose—not as masters of the universe, but as stewards of its beauty and complexity. The thread that connects us, from the origins of life to the dreams of the future, is the thread of curiosity. It is the force that drives us to seek, to create, and to grow. It is the thread that makes us human.

www.ingramcontent.com/pod-product-compliance
Lightning Source LLC
Chambersburg PA
CBHW052159220526
45471CB00004B/1743